Eternity Between Space and Time

Eternity Between Space and Time

From Consciousness to the Cosmos

Edited by
Ines Testoni, Fabio Scardigli, Andrea Toniolo
and Gabriele Gionti S.J.

DE GRUYTER

ISBN 978-3-11-131284-2
e-ISBN (PDF) 978-3-11-131361-0
e-ISBN (EPUB) 978-3-11-131408-2

Library of Congress Control Number: 2023950014

Bibliographic information published by the Deutsche Nationalbibliothek
The Deutsche Nationalbibliothek lists this publication in the Deutsche Nationalbibliografie; detailed bibliographic data are available on the internet at http://dnb.dnb.de.

Cover image: Valeriia Tretiakova/iStock/Getty Images Plus
Typesetting: Integra Software Services Pvt. Ltd.
Printing and binding: CPI books GmbH, Leck

www.degruyter.com

MIX
Papier | Fördert gute Waldnutzung
FSC® C083411

Contents

Abbreviations

AI	Artificial Intelligence
BB	Big Bang
BS	Breakdown of Symmetry
CCC	Conformal Cyclic Cosmology
CCR	Canonical Commutation Relations
CDM	Cold Dark Matter
CERN	Conseil Européen pour la Recherche Nucléaire
CMB	Cosmic Microwave Background
COBE	Cosmic Background Explorer
DDI	Dilaton-Driven Inflation
DT	Dignity Therapy
DWQ	Dipole Wave Quanta
EST	Eternity between Space and Time
FLRW	Friedmann, Lemaître, Robertson and Walker
GM	Newton's gravitational constant and mass of the object
GR	General Relativity
GUP	Generalized Uncertainty Principle
GUT	Grand Unified Theory
HBB	Hot Big Bang
IAU	International Astronomical Union
IPPP	Institutional Program on Psychiatry for the Person
K	Kelvin
LED	Light-Emitting Diode
LHC	Large Hadron Collider
NG	Nambu-Goldstone
OPT	Operational Probabilistic Theory
PBB	Pre Big Bang
PDI	Patient Dignity Inventory
PEM	Principle of the Excluded Middle
PNC	Principle of Non Contradiction
PST	Primary Structure of Truth
QED	Quantum Electrodynamics
QFT	Quantum Field Theory
QIP	Quantum Information-Based Panpsychism
QM	Quantum Mechanics
RU	Ricciardi and Umezawa
SBS	Spontaneous Breakdown of Symmetry
SM	Standard Model
SMI	Severe Mental Illness
SR	Special Relativity

https://doi.org/10.1515/9783111313610-203

VSED	Voluntary Stopping Eating and Drinking
WCH	Weyl Curvature Hypothesis
WHO	World Health Organization
WPA	World Psychiatric Association
ΛCDM	Lambda Cold Dark Matter

Ines Testoni, Fabio Scardigli, Andrea Toniolo
and Gabriele Gionti S.J.

Introduction

This book is titled *Eternity between Space and Time: From Consciousness to the Cosmos.* It is the outcome of three days of studies and discussions at an international conference held in May 2022 at the University of Padua during its 800th anniversary celebrations. Then, the title of the book is the same of the conference.[1] *Eternity between Space and Time* (EST) intends to challenge contemporary thought, untie a knot that bridles the entire history of human reflection and open up a new horizon of discussion about the relationship between infinite eternity and what appears finite, including consciousness. For over a century now, culture and academic research have established insurmountable boundaries between different fields of knowledge – thanks to and because of an increasingly rigorous and specialised methodology that differentiates the specificity of the objects of study in terms of philosophy and theology on the one hand and the hard sciences and physics, in particular, on the other. Between the aforementioned categories remains a reflection on the human condition, which is pushed in different directions at different times. Although the existence of contamination remains inevitable, such contaminations are not always highlighted. This book seeks to retrain the continuity of the same object of reflection and how it is the continuum within which any reasoning around the relationship between existence, reality and being gains its meaning even when the arguments seem strictly specialised and, therefore, incommensurable with respect to one another.

In fact, the concept of eternity is challenging because it appears to be exactly what it does not appear to be. However, EST intends to highlight how this concept supports the most rigorous investigations. The discussion is divided into the following four parts that include contributions from the keynote speakers in Padua: (1) "What about Eternity?", (2) "The Eternity Concealed in the Cosmos and the Secrets of Consciousness", (3) "Eternity, Time and Faith" and (4) "Existential Corollaries".

The first part gets to the heart of the issue of 'being' and specifically how the whole question of what is and what is not arises in it, i.e. what language endowed with meaning indicates. The question concerns whether *being* may not be and addresses the issue by referring to two metaphysical philosophers of contemporary thought, Emanuele Severino and Martin Heidegger, who have posed the question

1 https://www.unipd.it/news/eternity-between-space-and-time-consciousness-cosmos-est, last accessed 21 September 2023.

https://doi.org/10.1515/9783111313610-001

in a radical way. The connection between these two giant philosophers, who have reframed the importance of reflection on *being* by going back to the roots of Greek thought, had already been highlighted by the philosopher Massimo Cacciari on the occasion of the end of Severino's university teaching, who declared that the philosopher's lesson is not only "equal to that of Heidegger"[2] but also that there is an absolute opposition (aut/aut) between the two philosophers. The question was further considered in an international conference, "Heidegger nel pensiero di Severino" (Heidegger in Severino's Thought), held in Brescia in the year Severino himself passed away.[3] These two thinkers represent a continuity while also maintaining a distance between substantial components of continental thought developing between Germany and Italy. If Heidegger's contribution remains widely disseminated internationally in a vehicular language, Severino's contribution is beginning to be so through the translation of his three very significant works into English: *The Essence of Nihilism* (2015), *Law and Chance* (2023) and *Beyond Language* (2024).

Therefore, the more exquisite philosophical part of EST intentionally comprises its reflection on *being* and its eternity within this framework. In particular, in the chapter "The Eternity of Every Being and the 'Trace' of the Infinite in the Finite according to Severino", Giulio Goggi lays out the most specific feature of the fundamental ontology developed by Severino: the thesis according to which every being, qua being, is eternal. Then, the chapter will dwell on the topic of the 'trace' of the infinite in the finite as every being is eternal and necessarily stands in relation to every other being; it is necessary for every being to somehow be present in every other being.

In line with Goggi, Damiano Sacco's essay titled "Emanuele Severino. *Sózein tà Phainómena*" introduces some key elements of Severino's theoretical apparatus through a discussion of one of the key axes of the enquiries related to science and philosophy, which are epitomised by the tenet of saving (the appearing of) the phenomena (*sózein tà phainómena*). This standpoint affords an assessment of the radical and singular character of Severino's reflection as part of which the truth and eternity of every being appear as the impossibility for the being and appearing of every being to not always be saved.

In his article "The Absolute Appearing of Eternity as the Original Meaning of Time", Leonardo Messinese traces a further continuity between Severinus and Bergson based on the foundation of Greek thought. The author seems to dwell on the trait that unites the Platonic and Aristotelian conception of time and then on the critical analysis done by Henry Bergson. Subsequently, he compares the Berg-

2 Cacciari in *Corriere Della sera* and in *La Repubblica* (Cacciari 2001).

3 The conference was held in Brescia on 13–15 June 2019. The proceedings, edited by Ines Testoni and Giulio Goggi, are available here: https://www.padovauniversitypress.it/it/publications/9788869381577.

sonian reflections on time with those of Emanuele Severino to introduce the thesis that the absolute appearing of eternity is the original meaning of time.

In his article "Note on the Dialogue between Severino and Vitiello", Massimo Cacciari relates Severino with another Italian thinker, Vincenzo Vitiello, who has long dealt with Heidegger's thought and the entire continental tradition. The aforementioned comparison highlights some noteworthy basic ontological nodes.

Finally, the contribution of Roberto Tommasi, "Time, Eternity, Freedom in Kierkegaard, Heidegger and Ricoeur", investigates the relationships between space, time, freedom and eternity in Kierkegaard, Heidegger and Ricoeur. From the perspectives opened in this regard by the three thinkers emerges the aporetic oscillation between cosmological, existential and historical conceptions of space-time.

The second part of EST is titled "The Eternity Concealed in the Cosmos and the Secrets of Consciousness", and contains essays dedicated to the aspects of space and time that are intertwined with Physics and Consciousness. In particular, the essays of 't Hooft, Veneziano, and Penrose, explore the elusive concepts of time and eternity as they are conceived, on the one hand, in modern cosmological theories, and on the other, in those conceptual gymnasiums called black holes. In the latter, perhaps we begin to glimpse a profitable 'mixing', if not a unification, between the two great conceptual structures that still govern 21st-century physics, namely Quantum Theory and General Relativity. Scardigli's essay also follows this path by exploring the mix of concepts between gravitation and quantum indeterminacy. Instead, the contributions of Vitiello, D'Ariano, and Faggin appear almost as a counterpoint to these writings. Using the conceptual tools of today's theoretical physics, namely Quantum Information Theory, and Quantum Field Theory, the authors attempt an amazing exploration of the crucible where the very categories of space, time, reality and eternity are formed and built, i.e. (human) consciousness. These essays collectively provide the reader with 'windows' from which to glimpse unsuspected, perhaps astonishing panoramas that call for further journeys and explorations.

In his essay on the "Basic Ideas of Conformal Cyclic Cosmology" (CCC), Roger Penrose[4] illustrates his new vision (2005) of the cosmological theory. The CCC proposes that the universe undergoes repeated cycles of (accelerated) expansion, named 'aeons', where the maximal (or infinite) extension of the previous cycle goes to coincide with the Big Bang stage of the successive cycle. No contraction (big crunch) is required in this model. This is made possible through the conformal structure that dominates space-time at the beginning and at the end of each

---

4 It is important to point out that Roger Penrose had a discussion with Emanuele Severino at the conference organised by Fabio Scardigli at the Cariplo Congress Center (Milan) on 12 May 2018. The outcomes of the meeting are collected in: Penrose et al. 2022.

aeon. The CCC solves the paradox of the super-special initial conditions required by the Second Law at the Big Bang, and among its observational consequences, predicts the presence of 'circular rings' in the temperature fluctuations of the Cosmic Microwave Background spectrum.

In Gerard 't Hooft's[5] contribution titled "How Studying Black Hole Theory May Help Us to Quantize Gravity", black holes, far from appearing cosmic monsters or astrophysical curiosities, are instead described as the appropriate theoretical arena in which the basic principles of General Relativity uniquely intertwine with those of Quantum Theory. Therefore, it becomes possible to have a glimpse into the key roles that quantum effects play in gravitational interactions at ultra-short scales.

In his essay "Uncertainty Principle and Gravity", Fabio Scardigli describes how the uncertainty principle, the cornerstone of quantum mechanics, should be modified when gravity is properly taken into account. Among the many different physical predictions of this 'Generalized Uncertainty Principle', the possibility of considering black hole 'remnants' as sources of the enigmatic dark matter is briefly discussed.

Gabriele Veneziano's chapter "The Big Bang's New Clothes and Eternity" describes how the traditional role of the Big Bang is completely overturned in modern inflationary cosmology: the Big Bang is the instant at which the Universe, after having been cooled down to zero temperature, suddenly 'reheats' through an irreversible quantum process. As a consequence, the Hot Big Bang is associated with neither a singularity nor the beginning of time. It becomes therefore possible to enquire about whether time had a beginning, and how was the Universe before the Big Bang.

In his chapter "For a Science of Consciousness", Giacomo Mauro D'Ariano focuses on the topic of 'consciousness' or 'awareness'. He wants to ground 'consciousness' on either a physics theory or a physics theory-like base in order to bring a certain 'objectivity' to it. He claims that consciousness has a quantum nature and can be explained with quantum (interior) information theory. At the base of consciousness, there are q-bits (quantum bits). However, this interior information is subjective and cannot be transferred because the passage from interior (quantum) information to exterior (classical) information destroys interior information. Interior experiences are processed as quantum information. They are identified by the author with the 'qualia' of the philosophy of mind.

In his essay "Freedom and Artificial Intelligence", Federico Faggin tells that, after a mystical experience, he arrived at the conclusion that our universe is

5 It is important to emphasise that Gerard 't Hooft had a discussion with Emanuele Severino at the conference organised by Fabio Scardigli at the Cariplo Congress Center (Milan, Italy) on 13 May 2017. The outcomes of the meeting are collected in: Scardigli et al. 2019.

more than a materialistic reality as described by science. There exists the One, the totality of what exists. Consciousness and free will are part of the One and are described by a theory of quantum information. Consciousness is the inner space where signals from the external world are processed and become emotions, feelings and so forth. Free will is strictly connected to consciousness, it is the awareness that the experience I am having is my experience.

In his chapter "Brain, Mind, the Arrow of Time and Consciousness", Giuseppe Vitiello proposes to model the brain as a quantum field theory system. This system continuously interacts with its environment, and its functional activity is described by dissipative dynamics. The environment is described as a time-reversed copy of the brain called the Double. The act of consciousness inhabits the dialogue between the brain and its Double.

The third part, titled "Eternity, Time and Faith", is about theological–religious reflection.[6] In particular, it makes the biblical–Christian conception of time interact with the visions of time and reality proper to science and to modern and contemporary philosophy. The classical conception of physical-mechanical time has led to thinking of temporality (the condition of 'being in time') as a limit to be overcome and reach eternity (a condition in the future). According to this perspective, the meaning of human existence, subjected to time and the limits of transience and finiteness, emerges insufficiently. The understanding of temporality as becoming and limiting, as a lack of consistency and permanence, and therefore non-being, has negatively conditioned the very idea of revelation, or the way in which existence relates to the transcendent or hierophany – the manifestation of the sacred in human experience. The biblical resumption of time as an opening and place of revelation makes it possible to reshape the debate between science and faith (without confusion and separation) and to think of finiteness in close relationship with eternity and otherness as the revelation of the eternal.

Kurt Appel's contribution, "The Eighth Day. Biblical Time as Openness of Chronological Time" begins with the biblical creation story built according to a temporal narration. The seventh, or rather the eighth, day inscribes an openness

6 It is important to mention that there is ongoing work on the possibility of resuming the theological and theoretical discussion of Emanuele Severino's thought and Christian thought. The congress and this volume are part of this type of reflection that has been ongoing for some years now among scholars of theology and philosophy. We particularly highlight a webinar held on 24 June 2021, from 9.00 a.m. to 1.30 p.m. titled *Cristianesimo e Emanuele Severino. Quali possibilità di confronto? Approcci filosofici e teologici* (*Christianism and Emanuele Severino. Which possibilities for comparison? Philosophical and theological approaches*), the results of which are collected in a volume with the same title edited by Andrea Toniolo and Ines Testoni "Cristianesimo e Emanuele Severino. Quali possibilità di confronto? approcci filosofici e teologici" Padova University Press, available at: https://www.padovauniversitypress.it/it/publications/9788869382819.

in time that eludes all functionalisation. The *eschaton* (the seventh day) is the transition into the radical openness of time.

In his “The Time, Revelation or Negation of the Eternal? The Modern Metaphor of the ‘Death of God’”, Andrea Toniolo suggests that before the modern physical ‘revolution’ on the conception of time/space, it was the modern theological (and philosophical) thought that threw the ‘classical’ view of time and history into crisis. This crisis is emblematically expressed by the metaphor of ‘the death of God’ (Nietzsche, Hegel and Jüngel).

Piero Benvenuti’s chapter “Cosmology and Cosmologhia: A Much Needed Distinction” distinguishes, without separating, between the scientific models of cosmic evolution (cosmology) and the possible global cosmological models (*cosmologhia*). These models are anchored in scientific models; however, they differ by the choice of solution of the stumbling blocks encountered by scientific methods. They can be represented by the multiverse hypothesis, the cyclical universe or other philosophical or theological hypotheses.

In his “God and the Big Bang: Past and Modern Debates between Science and Theology”, Gabriele Gionti introduces the contemporary view on the birth (Big Bang) and evolution of our universe as well as the Hartle-Hawking model of quantum cosmology. He presents two models of the relationship between science and theology (and Church teaching) that occurred in history: (1) the ‘concordist’ view, since Big Bang theory appeared quite in agreement with Christian doctrine of creation and (2) the ‘complementary magisterial’ view, in which we distinguish between the scientific and theological planes as two parallel ‘lines’. To avoid confusion, it is necessary to regain a good conception of the doctrine of creation.

Alberto Peratoner’s contribution titled “‘Qu’est-ce qu’un homme, dans l’infini?’ Eternity and Infinity in Blaise Pascal and in the 17th-Century Geometrizing Ontologies” re-proposes the suggestive anthropological reflection of Pascal, who derives the human consciousness of his own condition from the *géométrie*, i.e. from the concept of infinity as a representation of reality that shows his condition as suspended between infinity and nothingness.

Finally, Leopoldo Sandonà, in his “Eternity and Otherness from the Perspective of Dialogic Thinking. Inspirations and Contaminations in and from Romano Guardini, Franz Rosenzweig and Nishida Kitarō”, approaches the relation between time and eternity from the innovative perspective of dialogic thought, crossing contemporary philosophy and theology with Jewish and Christian thinking. The eternity is not a concept but a relation, as Rosenzweig says, “the ‘us’ are eternal”.

The fourth part titled “Existential Corollaries” intends to reach the existential dimension of the human being, who thinks of eternity and totality in its ontological, physical and theological infinity and then finds himself having to come to terms with his own condition of finitude, searching for the arguments that can restore a substantial value and give meaning to life lived in experiencing different forms of pain and fatigue with which madness announces itself.

In her "Eternity, Instant, Duration. *Tangere aeternum*", Ilaria Malaguti considers how the centre of human existence, the actuality of the ego with itself, is enclosed in the intertwining of *chronos* and *kairos*. In our temporal and chronological experience, can we think of *kairos* as the instant in which we are offered the possibility of a *tangere aeternum*? Can we think of the moment starting from an interiority that does not withdraw into itself but becomes attentive and rises in intimate contact with the eternal?

Santo Di Nuovo's chapter titled "Finitude and Project: For Which Space? And for What Time?" reviews the challenges of finitude to philosophies, religions and sciences and reports the transhumanistic claim for artificially simulating an immortal consciousness. Based on some phenomenological suggestions and Edgar Morin's concepts of world citizenship and 'reliance', it presents some hypotheses for implementing a shared project of transcendence to begin in our present world.

Diego De Leo's chapter, titled "The Last Waltz: Finitude, Loneliness and Exiting from Life", discusses how the instrumentalist culture of modern society seems to have difficulty in dealing with the idea of life destined to end. Death seems to be considered for only old people. This chapter describes the problematic confrontation with finitude and unwanted travel companions in the course of life, such as loneliness, depression and suicidal ideation – conditions that make one wish for a different culture of death but, above all, a different preparation for life.

In their chapter "Beyond the Limits of Mental Illness: Dignity and Dignity Therapy in Person-Centred Psychiatry", Luigi Grassi and Harvey M. Chochinov consider how person-centred psychiatry and dignity-conserving care, including dignity therapy, should be practised in all mental health care settings to reduce the alienation, loss of identity, stigma and psychological, interpersonal, spiritual and existential suffering that people with psychiatric disorders have to face.

With her chapter "Beyond Alienation: Severino's Removal of Pathological Contradiction", Ines Testoni concludes the entire volume by bringing the whole discussion back to the opening discourse, that is to the Severinian ontological dimension that indicates the necessity of eternity. The substantial aim of this contribution is to highlight the inability to think of the eternal, i.e. how thought is still immersed in the radical madness of nihilism that consists precisely in thinking that being as becoming is nothing.

References

Cacciari, Massimo. 2001. “La sua lezione è pari a quella di Heidegger [His Lesson Is Equal to That of Heidegger]”. *La Repubblica*, 22 February. https://ricerca.repubblica.it/repubblica/archivio/repubblica/2001/02/22/la-sua-lezione-pari-quella-di-heidegger.html, last accessed 21 September 2023.

Penrose, Roger, Emanuele Severino, Fabio Scardigli, Ines Testoni, Giuseppe Vitiello, Giacomo Mauro D’Ariano, and Federico Faggin. 2022. *Artificial Intelligence versus Natural Intelligence*. Cham: Springer International. https://link.springer.com/book/10.1007/978-3-030-85480-5, last accessed 21 September 2023.

Scardigli, Fabio, Gerard ’t Hooft, Emanuele Severino, and Piero Coda. 2019. *Determinism and Free Will*”. Cham: Springer International. https://link.springer.com/book/10.1007/978-3-030-05505-9, last accessed 21 September 2023.

Severino, Emanuele. 2015. *The Essence of Nihilism*. Edited by Alessandro Carrera and Ines Testoni. New York/London: Verso [original: *Essenza del nichilismo*, Milan: Adelphi, 1982].

Severino, Emanuele. 2023. *Law and Chance*. Translated by Damiano Sacco and edited by Ines Testoni and Giulio Goggi. London/New York: Bloomsbury [original: *Legge e caso*, Milan: Adelphi, 1979].

Severino, Emanuele. 2024. *Beyond Language*. Translated by Damiano Sacco and edited by Ines Testoni and Giulio Goggi. London/New York: Bloomsbury [original: *Oltre il linguaggio*, Milan: Adelphi, 1992].

First Part: **What about Eternity?**

Giulio Goggi

The Eternity of Every Being and the "Trace" of the Infinite in the Finite According to Emanuele Severino

Abstract: In this article I will lay out the most specific feature of the ontology developed by philosopher Emanuele Severino: the thesis according to which every being, qua being, is eternal – a thesis founded on the *indisputable appearing of being in the form of identity/non-contradiction*. We shall see that the eternity of every being does not make the changing of beings illusory. Ultimately, the same foundation also underlies the inequality between what presently appears and the totality of beings, which Severino calls *infinite appearing*. I will then dwell on the topic of the "trace" of the Infinite in the finite: as each being *necessarily* stands in relation to every other being, it is necessary for each finite being to *somehow* include the totality of its "other".

1 Introduction

First of all, a terminological clarification is in order: what Severino means by "being" is anything that is not-nothing, e.g. a particular desk lamp, its ideal essence, the current state of the universe, the most fleeting of thoughts. The "being" of each of these determinations/differences signifies their not-being-nothing:

> That something "is" means primarily that it is not a Nothing, i.e., that it manages to keep to itself without dissolving into nothingness. And, in general, the plurality of modes of existence is nothing other than a plurality of the modes of not being nothing; so that the plurality of determinations or differences of Being is itself nothing other than the plurality of modes of existence, and any single determination is a unique mode of existence (Severino 2016a, 85–86).[1]

The thesis we will now be exploring runs as follows: it is impossible for anything that is, i.e. anything that is not-nothing, not to be, which is to say that it must necessarily be eternal. Here I will provide only an essential outline of the topic in question, and refer the reader to other publications for a broader discussion

1 Except for passages taken from *Essenza del Nihilismo* (*The Essence of Nihilism*, Verso 2016), the translations of excerpts quoted from other works by Severino are mine.

https://doi.org/10.1515/9783111313610-002

(Goggi 2019; 2022). Finally, I will touch upon the singular meaning of time implied by the eternity of every being, as well as some significant implications concerning the relationship between the finite and the Infinite.

2 The Eternity of Every Being

1. To think that any given being could not have existed and that it could cease to exist is to envisage a time in which this given being (this not-nothing) is nothing, absolutely nothing. But *since it is impossible for the non-identical to be identical,* not only is it impossible for nothingness to belong to any being *when* (i.e. *for as long as*) that particular being exists, but it is *absolutely* impossible for nothing to belong to it. The foundation of the thesis of the eternity of every being *qua being* is the necessity that each being be self-identical, i.e. the impossibility that any given being be other than itself: since the identification of non-identical meanings is absurd, and since what is absurd is what cannot be, it is impossible for any being not to be. And this means that every being, *qua being*, is eternal.

2. Severino puts it as follows: the law of being is the law of the *opposition between the positive and the negative.* Note that what he means by "positive" is every being, every not-nothing, whereas by "negative" he means anything that is "other" with respect to the positive under consideration. For instance, if we take "this lamp", its negative is not only any other positive which is not "this lamp": what is also other than "this lamp" is "nothing" – not in the sense that "nothing", i.e. the absolute absence of any positiveness, is in itself something, a being, but in the sense that "this lamp" is not nothing. Now, the need to affirm *the opposition between the positive and the negative* involves the affirmation of the eternity of every being because it involves that *specific opposition between the positive and the negative* that consists precisely in this, namely that every being is not nothing:

> It is *necessary* to affirm that every being is eternal, because eternity is *one* opposition between the positive and the negative (it is that opposition by which the positive, any given being, is not nothing), which is to say that it is a form, a specific mode of that – the universal opposition between the positive and the negative, the universal determination of the being – the negation of which coincides with self-negation. The necessity of affirming the opposition between the positive and the negative [which is inclusive of every specific form of this opposition] *necessarily* implies the affirmation of that specific opposition between the positive and the negative which is the eternity of every being (Severino 1995, 243–244).

Severino has called this "the golden implication", on account of its remarkable significance. It should be noted that in claiming that *everything is eternal,* we are not saying that everything exists according to a particular mode of being, for example the mode in which spatio-temporal beings exist; rather, we are saying that every not-nothing is eternal, that is *every mode of being* (whether it be spatio-temporal, ideal, fictional, obscure . . .). Therefore, it is necessary to state that *every being*, i.e. *every meaningful being*, is not nothing and that it is impossible for it to become nothing or to have been nothing, since this becoming nothing and having been nothing implies the contradictory identification of non-identical meanings.

3 The Indisputability of the Opposition between the Positive and the Negative

1. Leibniz wondered why something exists, rather than nothing. This has gone down in the history of philosophy as the "fundamental metaphysical question". But according to Severino this question leaves open the contradictory assumption that something (i.e. beings) could not exist, whereas beings *do exist*, for it would be contradictory for them not to.

2. If it is crucial to envisage the eternity of every being, founded on the necessity that each being be self-identical and other from what is other than itself, what is equally crucial is to show that this opposition between the positive and the negative is undeniable. Severino proves it via "refutation", by developing – arguably like no other philosopher before him – the *elenchtic* strategy that Aristotle has laid out in *Metaphysics*, Book IV. I will sum it up as follows: the negation of the difference of differents, however it presents itself, presupposes the appearing of the difference of differents; for if differents did not appear as differents, no negation of difference would emerge; but this means that, in negating the difference of differents, this negation negates its own foundation, i.e. what constitutes it (namely, what enables it to exist as a negation), and hence negates itself. Severino writes:

> In order to have a real negation of the opposition (and not merely an apparent one), it is necessary that the positive and the negative should first be posited as different (and so as opposites), and that one then posit the identity of the differents, i.e., that the differents *qua* differents are identical. As long as the differents are not seen as different, they must unquestionably be said to be identical; but if they are seen as different, and if, indeed, they must be held fast as different, in order that the affirmation of their identity may be negation of the opposition of the positive and the negative, then this negation is grounded upon the affirmation of what it denies; and, this time, it is no longer grounded upon the affirmation of only a part of what it denies, but rather upon the whole content that is denied. Consequently, the negation is negation of that without which it cannot constitute itself as nega-

> tion, and so is negation of itself; it is a quitting the scene of the word and of thought, a declaring its own nonexistence and its own meaninglessness (Severino 2016a, 69–70).

The negation of the difference of differents removes itself, *and it is precisely this essential self-removal that makes it necessary for every being to show itself in the form of identity/non-contradiction*, which implies the affirmation of the eternity of being as such.

4 The Singular Meaning of Becoming and Time

1. If the existence of time implies the existence of a "before" and an "after", and if "before" and "after" are understood as the fluctuating of things between being and non-being, then time is non-existent: what exists is the belief that time exists, but this is "the time of the absurd" (Severino 2016a, 88), something that cannot exist and which therefore cannot be attested by experience. Let me better explain this point.

2. A body burns and is replaced by ash. What is it that appears to someone witnessing this process? Does it appear that the body has become nothing? Does its annihilation appear? Severino writes:

> After the fire, ashes; which means: when the fire no longer appears, ashes appear. But that something that no longer *appears* no longer *is* – *this* is not manifest in Appearing. On the contrary – it is *interpreted* on the basis of the way in which something appears and disappears. When something appears that has never appeared before, one says that it has been born and that previously it was a Nothing; when something disappears and does not return, one says that it has died and become a Nothing. And men have learned that when something appears in a certain way, it has never appeared before; and when it disappears in a certain way, it will not return (Severino 2016a, 109).

Science says that the amount of matter in the universe remains constant, even though it takes different forms: the energy currently found in the universe was already present at the time of the Big Bang, but it was concentrated in a tiny volume. Let us consider the process whereby wood turns to ash and ask ourselves: "Before the ash was produced, did it already exist? And once the wood turns to ash, will it continue to exist?" Well, insofar as the wood and the ash are a certain amount of energy, they do not become nothing and do not emerge out of nothing. But what happens to the wood *qua wood* – i.e. to that specific form we call wood – when it turns to ash? And what about the ash *qua ash* – i.e. that specific form we call ash – before it is produced? Science and the whole of Western thought tell us that the wood (qua wood) no longer exists when it turns to ash and that the ash (qua ash) did not exist yet before it was produced: if this were not the case, there would not be any becoming. But are things really so? Consider, first of all, the fact that appearing does not reveal this "no longer existing" and "not existing yet" *in any way*:

> When the wood (qua wood) has becoming nothing, does it continue to be observable, experienceable, ascertainable? [. . .] Certainly, if one believes that things become nothing, one must believe that insofar as they become nothing, they are no longer observable, experienceable, ascertainable as they were before. [. . .] To experience is to experience an existent: it is impossible to experience what is now nothing (Severino 2015a, 188–190).

A similar argument must be made for ash: if one believes that the ash (qua ash) was nothing before it was produced, then, insofar as it was nothing, it could not be part of the totality of what is experienced: for one cannot experience nothingness. But this means, precisely, that it is impossible for experience to say anything about the fate of that which is believed to have gone into nothingness or to still be nothing. Certainly, there are certain modes of becoming in relation to which man has convinced himself that certain things have emerged out of nothingness and will return to nothingness:

> In relation to many things, including many that are dear to him, [man] experiences that, when they no longer show themselves with the traits they used to display, they no longer return. [. . .]. And [it happens that] in relation to those things that are born one goes so far as to say that they have emerged out of nothing, because they have never been seen before: as if someone who witnesses this birth had the capacity to experience the infinite times past [. . .], and thus to discern that what was "born" in them just wasn't there, never has been there, i.e. was nothing (Severino 2015a, 191–192).

But since it is impossible to experience nothingness, and hence to experience annihilation, stating that things are born and die, that they are generated and perish, is an interpretation which alters what genuinely appears:

> This means that becoming other is the content of a *theory* established on the basis of the delusion caused by the non-return of what no longer appears (but what human being has ever experienced the infinity of future times [. . .] so as to be able to claim that what has faded will never return?) (Severino 2015a, 192).

But what, then, is that which genuinely appears? If (and because) every being is eternal, the varying of things and situations we experience cannot coincide with the coming to be or ceasing to be of beings; rather, it must be the supervening of eternal beings within the eternal horizon of appearing:

> The Becoming that appears is not the birth and the death of Being, but rather its appearing and disappearing. Becoming is the process of the revelation of the immutable (Severino 2016a, 111–112).

Not even that appearing which begins to appear and ceases to appear can be something that begins to be and ceases to be: when something appears, its appearing necessarily appears (for otherwise what would appear would be something that does not appear). It follows that when something begins to appear, its beginning to appear also begins to appear: within the total horizon of appearing – what Severino calls "transcendental appearing" – the appearing of something begins to appear (not: begins to be!); and when something ceases to appear, from the total horizon of appearing the same appearing of something ceases to appear (not: ceases to be!). Becoming occurs when eternal beings (and their eternal appearing) enter or exit the stable transcendental dimension of appearing; however, the supervening of this dimension and its departing from appearing cannot "appear": for the appearing of becoming is only possible if this appearing is not becoming, but rather the unchanging background that encompasses the totality of time, which is to say every "before" and "after" that begin to appear.

3. As it is necessary for what begins or ceases to appear to also *be* before it begins to appear and after it ceases to appear – for every being is eternal – we will say that present beings are eternal, but so are past and future ones:

> This day *is* (eternal), even when what now appears as the past was the present and when what now appears as the future will be the present; in turn, past and future beings *are* (eternal), in the concreteness that pertains to them when they have been and will be the present, even when this day appears. *If* this concreteness of theirs differs from what appears of them when this day appears [. . .] this means that, in the past and future appearing together with this day, this concreteness of theirs has (respectively) disappeared and not yet appeared (Severino 2015b, 139).

In Italy some scholars (Soncini and Munari 1996) have sought to compare the "Parmenidean" Einstein to Severino and the thesis of the eternity of every being. As is widely known, Einstein's special theory of relativity leads to the remarkable conclusion that all things within space-time – things past, present, and future – are eternal. However, Severino himself noted that the *necessity* that his writings bring into play is something essentially different from the *hypothetical-deductive logic* underlying scientific demonstrations. As we have seen, the eternity of every being *qua being* is a specification of the impossibility for anything (i.e. any being) to be other than itself. Severino speaks of the "originary structure" of knowledge to refer to this fundamental and indisputable appearing of every being's self-identity, which is far from hypothetical and implies the eternity of every being:

not just the eternity of those beings that belong to the space-time dimension, but also the eternity of the *non-spatial* dimension of every being. What is eternal is not only every configuration of the world, but also every state of consciousness, every emotion, and every concept. Furthermore, according to Einstein – as for Parmenides before him – the experience of change is *illusory* because it shows beings passing from non-being into being: in his famous *Letter to the Family of Michele Besso*, Einstein wrote that "the distinction between past, present, and future is only a stubbornly persistent illusion". According to Severino, by contrast, experience attests to variation, but not to the transition from non-being; hence, it is not at all illusory, and nor is time, understood as the supervening of eternals, their appearing and disappearing within the everlasting horizon of appearing:

> Every being is at all times, in the sense that although it does not appear at all times, it coexists with what progressively appears in time, which is to say at all times (Severino 2015b, 140).

Severino is the philosopher of the eternity of every being, but he is also the philosopher of time understood as the coming forth of eternals. He has called this coming forth of eternals "Glory", showing that it is destined to continue forever. And since every being is eternal, every being "is" even before its appearing, and continues to "be" even after it has disappeared. So the totality of what presently appears cannot be the dimension of the totality of beings, which leaves nothing outside itself. Severino calls it "infinite appearing", pointing out that a totality which did not appear to itself (i.e. that lacked its appearing) would not be the totality of all beings.

5 The "Glory" of Every Being

1. The originary structure of being is the essential predicate of every being. It represents a set of interrelated meanings (being, nothingness, appearing, identity, difference . . .) that is untranscendable, in the sense that it constitutes what lies in the background of all appearing: no being could appear if it did not appear as what is identical to itself and other from what is other with respect to it. Now, any supervening thing that were untranscendable – just as the background of all appearing is untranscendable – would be something that (insofar as it is supervening) begins to be united necessarily with the determinations of the background. But this beginning to be implies the absurd, i.e. the (initial) nullity of that being in which this union consists. Therefore, it is impossible for any supervening thing to interrupt the spectacle of the supervening of eternals:

> an untranscendable supervening thing is impossible and self-contradictory, insofar as it implies the nullity of being (and, strictly speaking, implies the nullity of itself [. . .]). Therefore, any supervening thing is necessarily transcended; and since any transcending thing is a supervening one, it is necessary for the supervening [of beings] to unfold infinitely (Severino 2007, 237).

The fundamental meaning of this Glory – which is ultimately "the genuine meaning of time" (Severino 2007, 205) – is the infinite unfolding of beings within the finite circle of appearing; and since the beings destined to supervene are infinite, so must be the beings destined not to supervene. The totality of this content must belong to the infinite appearing of beings, which is infinite insofar as nothing appears beyond it and the beings appearing within it are infinite.

2. The Glory theorem – i.e. the claim that it is impossible for what supervenes in the originary circle of appearing to be something untranscendable – also implies the existence of an infinite multiplicity of finite circles of appearing. Indeed, the actuality of the supervenient (i.e. the actual appearing of what supervenes in the transcendental horizon of appearing) is itself something which supervenes, and hence cannot be untranscendable either. In this case, what we have is the necessary transcending of the *actual appearing* that pertains to that which supervenes, insofar as it appears within the originary circle of appearing; and this transcending can only be the supervening of beings within a *different* dimension of actuality from that which pertains to the originary circle of appearing:

> This different and transcending actuality, in other words, supervenes in *another circle of finite appearing* (Severino 2001, 172).

And since the appearing of what supervenes in this second circle is itself supervenient, it must be argued that it too is transcended by what supervenes in a third circle, and so on, *in indefinitum*. What are infinite, therefore, are the finite circles of appearing: those dimensions within which the originary structure of the truth of being has always shown itself (and will always show itself), i.e. the appearing of the being-itself of every being and its implications.

3. In relation to infinite appearing (which is the dimension of the totality of beings) there is no supervening or disappearing, in the sense that nothing enters or exits it. But this does not disprove the totality of the supervening that appears in the infinite finite circles of appearing. It may be argued, instead, that within infinite appearing that supervening appears in the totality of its unfolding: within it something eternal appears, namely the infinite unfolding of those beings destined to make their way into the infinite finite circles of appearing and much more besides – infinitely more. This has nothing to do with any kind of theological transcendence: Augustine, Aquinas, and any creationist perspective are over-

come. The finite which appears in infinite appearing is that finite which appears here and now: in finite appearing it shows itself in an abstract way, whereas infinite appearing is the very totality of the positive in its semantic concreteness; therefore, it coincides with the surpassing of the finite and hence too of finite appearing (and of the totality of the contradictions of the finite), where being appears in a processual way.

6 The "Trace" of the Infinite in the Finite

1. The finite is a contradiction not because, as Hegel assumed, it becomes other than itself (for "becoming other" is impossible), but rather because it necessarily stands in relation to every other being and to the totality of beings, which do not appear in their concreteness. We have seen how the originary structure of knowledge is also a finite horizon:

> Since [the] originary [meaning] is and means what it is and means only in its connection with the All [. . .], in the isolation of the originary from the All (i.e. in the non-manifesting itself of the All in the originary), the originary is not the originary (Severino 1981, 73).

The contradiction of the finite is determined by the abstractness of its position – whereby what is posited is not what one intends to posit – whose removal is given not by the negation of its content, but by the concrete manifestation of the All, which is precisely the appearing of the totality of beings. Bearing in mind the conclusion we have reached through the Glory theorem – namely, the need for supervening beings to unfold infinitely, by supervening within the transcendental horizon of appearing – we may argue that the transcending of the contradiction of the finite

> is an infinite journey, an indefinite expanding of the finite circle, such that the contradiction of the finite, qua finite, infinitely endures in its being infinitely transcended (Severino 2005, 89).

It may thus be argued that infinite appearing coincides with the eternal transcending of the totality of the contradictions of the finite. Every being therefore belongs to the totality of beings, as does the infinite unfolding of beings destined to supervene in the infinite finite circles of appearing. We shall now see in what sense it is necessary to argue that in every being the infinite totality of beings appears: not only in the generic sense whereby we say that the appearing of X is the appearing of its being other with respect to all that is other than X – so that the analysis of X generically reveals the formal meaning of the whole – but also in a far more specific sense.

2. As every being is eternal, each being *necessarily* stands in relation to every other being; and since this relation is necessary, it is necessary for every being to be *present* in each individual being. Let X and Y be two beings: if X in no way appeared in Y (and if Y in no way appeared in X), there would be no relationship between X and Y, whereas the eternity of every being implies that X and Y are *necessarily* related each other (and to every other being). And if there was no relationship between X and Y, then neither X would be other than Y, nor Y would be other than X, i.e. neither X nor Y would be themselves. But how is X present in what is other than X? First of all, Severino notes that this presence is possible insofar as what is present is not the other in its concrete determinateness:

> It is *necessary* that any being X – in and in relation to its concrete determinateness – be nothing *in any other being* Y, and that the concrete determinateness of Y (i.e. of any other being) be nothing in X. The concrete determinateness of X in Y is nothing. [. . .] In other words, it is necessary that something in X be nothing in Y and vice-versa: for otherwise X would be Y (Severino 2015b, 142).

In Y the "abstract form" of what is other than Y will thus be present (likewise, the "abstract form" of what is other than X will be present in X). And the abstract form of X, which is present in Y,

> is not separate from the concreteness of X [. . .]; in fact, it is the 'representative', the 'spokesman' of that concreteness (Severino 2016b, 181).

Severino calls this presence "trace" and argues that it is a kind of *inclusion*: X, as the trace of X, is present in Y, and this is not a contradiction only insofar as X is present as what is negated, given that Y is not (i.e. does not mean) what is other than itself. In every being we must therefore distinguish between its *concrete part* (whereby it differs from its other) and its *abstract part*, which is precisely the presence in it – as what is negated – of the abstract form of its other:

> The fact that X, *qua* X, is present, in Y, as what is negated means [. . .] that X qua X, in Y, is nothing; but the fact that the abstract form of X is present, in Y, as what is negated does not mean that, in Y, it is nothing; rather, it means that Y is, in itself, the negation of this form, which in turn is a being. This means that, given the abstract form X's belonging to Y, the negation of this form (the fact of not being identical to this form) is not proper to Y qua Y, but rather to that *part* of Y that is not such a form (but it is proper to Y insofar as it includes such a part): it is that *concrete* part of Y which, being itself, is the negation of all that in Y is the *abstract* form of all that is not Y. The abstract form is the abstract part of Y. It is by virtue of its concrete part that Y is Y. If this concrete part did not exist, what would exist would not be Y, but the abstract form of all that is not Y (Severino 2016b, 189–190).

In those cases in which the relationship between X and Y is such that X is part of Y not only because there is a necessary relationship between X and Y, but also because X and Y are configured in a certain way – e.g. when X is "this lamp" and Y is the totality of beings of which "this lamp" is part – the concrete determinateness of X is also contained in Y, without thereby disproving what has previously been argued:

> The part exists in the whole, as that whose existence is affirmed; yet, the part is not the whole and therefore in some sense, or according to some aspect, "being a part" is nothing in "being the whole" and vice-versa; and the abstract form of "being a part" is – as what is negated – in "being the whole". "And vice-versa", meaning: just as the whole includes – as what is negated – the abstract form of the part, so the part includes – as what is negated – the abstract form of the whole (Severino 2015b, 146).

Moreover, if it is true that each being is made up of its (*concrete* and *abstract*) parts, and that none of a being's parts is the being itself, it is equally true that no being can be reduced to the sum of its parts isolated from each other. What makes a being a "totality" is the relationship between its parts, and since every being is eternal, what is also eternal (and hence necessary) is the relationship between its parts:

> Every totality ["not just those totalities that are in turn parts but also totality *simpliciter*, which is not a part of anything"] is the unity and relationship between its parts. [. . .]. But a totality is not the mere set – the mere "sum" – of its parts, for a totality is the eternal and necessary relationship between them (a specific relationship, which distinguishes each totality from all others): it is "constituted" by this relationship; and this means that such a totality is itself. Nonetheless, the fact remains that even though the parts of a being are bound by an eternal and necessary relationship, none of them are the being itself (Severino 2011, 261–262).

Ultimately, the necessity that each being (including that being we call the appearing of beings) should stand in relation to every other being (and to the totality of beings) implies the necessity that every being eternally include – as we have seen – the traces of all other beings:

> In the appearing of the sound of rain, the trace appears of sunshine, the sky, the most distant galaxies [. . .] every other being. [. . .]. In the appearing of the most irrelevant part of the Whole, the infinite traces of every other being appear (Severino 2001, 223–224).

Every being echoes with an infinite multiplicity of sounds, a kind of infinite symphony: these are traces of the infinite totality of beings and hence also of the infinite finite circles of appearing, and of the infinite unfolding of Glory itself: such traces are necessarily present. And while within the finite horizon of appearing – where the concreteness of beings appears in a processual way – the Infinite as such cannot appear (for the finite cannot become the always having been the infinite appearing of the whole), the affirmation of the necessary existence of the Infinite already constitutes a trace, in the finite, of that Infinite which is the appearing of the concrete totality of the infinite (eternal) relationships of each being to every other being and to this infinite totality of relationships.

References

Goggi, Giulio. 2019. "Golden Implication. The Primary Foundations of the Eternity of Being". *Eternity and Contradiction*, Sept. 2019: 43–56. https://ojs.pensamultimedia.it/index.php/eandc/issue/view/206/56, last accessed 19 September 2023.

Goggi, Giulio. 2022. "The Identity and Eternity of Every Being". *Eternity and Contradiction*, Dec. 2022: 80–99. https://ojs.pensamultimedia.it/index.php/eandc/issue/view/302/207, last accessed 19 September 2023.

Severino, Emanuele. 1981. *La struttura originaria* [*The Originary Structure*]. Milan: Adelphi [First Edition (1958). Brescia: La Scuola].

Severino, Emanuele. 1995. *Tautótēs* [*Identity*]. Milan: Adelphi.

Severino, Emanuele. 2001. *La Gloria* [*The Glory*]. Milan: Adelphi.

Severino, Emanuele. 2005. *Fondamento della contraddizione* [*Foundation of the Contradiction*]. Milan: Adelphi.

Severino, Emanuele. 2007. *Oltrepassare* [*Passing Beyond*]. Milan: Adelphi.

Severino, Emanuele. 2011. *La morte e la terra* [*The Death and the Earth*]. Milan: Adelphi.

Severino, Emanuele. 2015a. *In viaggio con Leopardi. La partita sul destino dell'uomo* [*Traveling with Leopardi. The Match on the Fate of Man*]. Milan: Rizzoli.

Severino, Emanuele. 2015b. *Dike* [*Justice*]. Milan: Adelphi.

Severino, Emanuele. 2016a. *The Essence of Nihilism*. Edited by Ines Testoni and Alessandro Carrera. Translated by G. Donis. London/New York: Verso [Originally published in Italian by Paideia as *Essenza del nichilismo* (1972). Brescia: Paideia. Second Edition (1982). Milan: Adelphi].

Severino, Emanuele. 2016b. *Storia, Gioia* [*History, Joy*]. Milan: Adelphi.

Soncini, Umberto, and Munari, Tiziano. 1996. *La totalità e il frammento. Neoparmenidismo e relatività einsteiniana* [*The Totality and the Fragment. Neoparmenidism and Einsteinian Relativity*]. Padua: Il Poligrafo.

Damiano Sacco

Emanuele Severino. *Sózein tà Phainómena*

Abstract: The principle of 'saving the phenomena' appears to structure the history of physics as much as the history of philosophy or meta-physics. Emanuele Severino's singular philosophical contribution may be traced to an unprecedented answer to the meta-physical question of saving the phenomena: phenomena do not need to be saved, Severino argues; they have always and already saved themselves. Their having always and forever saved themselves consists, according to Severino, in their eternity.

According to what we know from Simplicius, one of the earliest characterisations of the aim of the enquiry of 'physics' may be attributed to Plato himself, and may be condensed into the formula: *sózein tà phainómena* – to save the phenomena. Strictly speaking, however, it is not 'physicists' (*phusikoí*) who aim to 'save the phenomena', but 'astronomers' (*astrológoi*, or *astronómoi*). For, indeed, *phusikoí* enquire into the nature and the essence of *tà phusikà* (i.e. into the nature or essence of nature itself, *phúsis*) – thus carrying out an analysis that we would today call 'meta-physical' – whereas the enquiry of *astronómoi* resembles our modern science of physics. *Astronómoi* bear witness to the motion of the celestial bodies, and attempt 'to save what appears': 'to save the phenomena'. In his commentary to Aristotle's *De Caelo*, Simplicius writes: "The associates of Eudoxus and Callippus and those up until Aristotle hypothesised counter-revolving spheres homocentric with the universe and tried to save the phenomena by means of them" (Simplicius 2011, 65); "Astronomers [*astronómoi*] assume certain hypotheses and save the phenomena, agreeing that all heavenly things move uniformly" (Simplicius 2004, 80); "If it is possible to save the same things if the principles are hypothesised to be finite and if they are hypothesised to be infinite, it is better, Aristotle says, to take them to be finite (as also in mathematics) and as few as possible" (Simplicius 2009a, 84). However, what does 'saving the phenomena' mean? Simplicius writes:

> Perhaps the Pythagoreans and Plato did not hypothesise that the construction from such triangles was certainly like this in every respect, but rather they did so in the way that astronomers make certain hypotheses, different ones making different hypotheses and not insisting that the variegation in the heavens is certainly like this but that *when principles of this kind are hypothesised the phenomena can be saved* [*sózesthai tà phainómena*] with all the heavenly bodies moving in a circle in a uniform way (Simplicius 2009a, 39–40).

https://doi.org/10.1515/9783111313610-003

'To save the phenomena' thus means to put forth certain hypotheses (concerning, for instance, the phenomenon of the motion of celestial bodies) that are consistent with what actually appears. Through these hypotheses, the appearing of the phenomena is 'saved' (*sózesthai*). Astronomers are able to save the appearance of the retrograde motions of planets through specific hypotheses concerning their uniform circular motions: that is to say, the appearing retrograde motions of planets may be accounted for by a combination of uniform circular motions.

It may be argued that the history of physics unfolds according to the guiding principle of *sózein tà phainómena*. In the modern era, Kepler's hypothesis of an elliptic motion of planets appears to 'save' the motion of planets to a higher degree of precision than the one afforded by the Aristotelian doctrine; Newton's hypothesis of a universal gravitational force appears to widen the scope of phenomena that may be 'saved'; Einstein's hypothesis of a space-time manifold appears to be able to save phenomena that could not be accounted for by Newton's laws (and, in fact, physics goes to a greater and greater length to produce and observe the very appearing of the phenomena that are to be saved).

The crucial remark to be made, however, is that – as Simplicius already notes – the principle of 'saving the phenomena' does not only guide the enquiry of physics (the enquiry of 'astronomers'), but also the enquiry of those who enquire into the nature and essence of *tà phusiká* (and, later, of *tà metà tà phusiká*): i.e. philosophers.[1] Duhem's seminal 'hypothesis' (Duhem 1969), according to which the principle of 'saving the phenomena' guides (at least part of) the history of physics (a hypothesis devised precisely in order to save the appearing of the phenomenon constituted by the 'history of physics') should thus come to be extended to include the very history of philosophy or metaphysics (thus leading to the conclusion that philosophy in the West has been but a physics). For, in the same way in which astronomers or physicists put forth certain hypotheses concerning the planets that are able to reproduce their appearing motions, 'philosophers' have, throughout the history of the West, hypothesised different meta-physical realities that are able to save the appearing of *all* phenomena (and that are meta-physical in that they are not part of what immediately appear as part of the world of *phúsis*).[2]

1 "Plato makes clear that these things [namely, the figures of the *Timaeus*] are like the hypotheses used by the astronomers with which, when they are hypothesised, it is possible to preserve the phenomena" (Simplicius 2009b, 32).

2 As it will appear, the question of the relationship between what appears and the hypothesis that saves it well exceeds every question of whether a distinction between what is real and what appears exists for the Greeks, whether a distinction obtains between ontology and epistemology, what the status of Plato's 'likely' account in the *Timaeus* is, etc.

It appears that all phenomena, precisely insofar as they are phenomena, share certain properties – properties that, themselves, appear. That is to say, all phenomena, *qua* phenomena, appear insofar as there appears a certain primary or base content. This primary content is the appearing content that is shared by every phenomenon *qua* phenomenon (i.e. by every being *qua* being). Philosophy has aimed to advance different hypotheses that would save the appearing of this primary content.[3] These hypotheses include the *archaí* of the *phusikoí*, the atoms of the atomists, Parmenides' being, Plato's ideas, Aristotle's *ousía*, Plotinus' One – all the way to Spinoza's substance, Leibniz's monads and Kant's thing in-itself. Starting with Hegel, however, philosophy appears to realise that what, in this way, is posited beyond the appearing of phenomena (i.e. what is posited beyond, *metà*, the appearing of *phúsis*), and is argued to be able to save the phenomenon – *itself* appears: that is to say, German idealism, and primarily Hegel, brings to the fore the intrinsic contradiction of a meta-physical dimension that would, precisely, be beyond (*metà*) the appearing of the phenomenon, and that, *at the same time*, at least in some respect, would have itself to appear (for, otherwise, nothing could be thought or known of it).[4]

The contradiction of the principle of *sózein tà phainómena* thus lies in the notion of a 'hypothesis' – i.e. something *posited* (-thesis) *below* (hypo-) – that sustains, and *saves*, the appearing of the phenomenon *while not being itself sustained or saved*. The appearing of the phenomenon may be 'saved' only insofar as what saves it is, in turn, not saved. Starting with Hegel, philosophy recognises this contradiction, and yet, appears to be unable to overcome it. Contemporary philosophy may thus only posit the very contradictory character of the project of saving the phenomena: either in the contradictory form of an ungrounded principle (e.g. the Heideggerian *Abgrund*), or in that of the infinite *regressus* or deferral of the very principle that should be able to save the appearing of the phenomenon (Derrida's trace consisting precisely in this infinitely deferred and ungrounded principle).[5] And yet, the phenomenon *does* appear – and so does the contradictory

3 This is the case even when, for instance with Parmenides, the appearing content itself is argued to be an illusion: philosophy must then be able to save the appearing of the very illusory character of everything that appears.

4 This is the case despite the fact that, e.g. in Kant, there appears a difference between the possibility of thinking the thing in-itself and the possibility of knowing it: the thing in-itself may not be known, but its existence, identity and non-contradictory character may be thought.

5 This is the case for what concerns 'continental' philosophy; 'analytic' philosophy, on the other hand, appears unable to altogether detect this (i.e. its own) contradiction, and still pursues various attempts at saving phenomena (presently, for the most part, through the 'hypothesis' of language – i.e. believing that the analysis of language may save the appearing of linguistic phenomena, thus leaving the appearing of the very phenomenon of language 'un-saved').

character of the project of saving its appearing. How may the phenomenon appear, if its very appearing may not in any way be saved? How may the very project of saving the phenomena appear, if it is but an impossible and contradictory endeavour? Emanuele Severino's reflection appears precisely at this juncture.

1 Eternal Salvation

According to Emanuele Severino, the contradictory content of a 'hypothesised' (i.e. hypostatised) being that would save the phenomena – i.e. the contradictory content of the project of the history of philosophy – may *not* altogether appear: precisely because the content of a contradiction (the self-contradictory meaning) is nothing, and nothingness itself (the *nihil absolutum*) may neither be nor appear. Accordingly, the contradiction of saving the appearing of phenomena must in fact be negated: phenomena *cannot* be saved – but not insofar as they are always and irremediably sinking into the abyss, or insofar as the project of saving them consists of an infinite task, progression or deferral. Phenomena may not be saved in that they *do not need to be saved*: the very project of saving them is nothing. That phenomena do not need to be saved means that they *immediately* coincide with their own hypo-thesis: that is, phenomena immediately coincide with the very ground that saves their appearing. Emanuele Severino's enquiry consists, in the first place, precisely in a study of the (so-called "originary") structure of this immediacy: the immediacy of the identification of a phenomenon and its *phaínesthai*, of a being and its being, of what is grounded and its ground ("The originary structure is the essence of ground. In this sense, it is the anapodictic *structure* of knowledge – the *archè tês gnóseos* – the *self-structuring* of immediacy itself"; Severino 1981, 107).[6] The investigation into the structure of immediacy leads Severino to assert the nullity of the project of saving the phenomena (be that project part of the enquiry of philosophy, physics, science, technology, etc.). That phenomena do not need to be saved means that they always, immediately and forever save themselves: they immediately coincide with their own self-identity, and they are immediately saved by their coincidence with the ground of their self-being and self-identity. Crucially, according to Severino, this immediate self-being and self-identity, through which beings immediately and forever save themselves, constitutes their *eternity*. Every being and every phenomenon is im-

6 All translations from the Italian are mine unless stated otherwise.

mediately saved in that it is eternal: every being is first of all saved from the impossible annihilation that its becoming would entail.[7]

The guiding principle of saving the phenomena nevertheless appears to structure both the history of physics and the history of philosophy (despite the fact that the goal of this project – grounding the appearing of phenomena – is in any case impossible and self-contradictory: for, once again, according to Severino, phenomena may only be saved by their own eternity). According to Severino, the histories of physics and metaphysics, as guided by the project of saving the phenomena, may however be divided into two fundamentally different stages:[8]

(I.) In the first stage, philosophy (together with physics and the other abstract sciences) posits, beyond the domain of phenomena, a ground that saves the appearing of phenomena themselves: that is to say, philosophy posits a dimension of being that exceeds the present appearing of the phenomenon and constitutes its ground and its cause. This dimension, insofar as it lies beyond the manifest becoming of the world, is 'immutable'. This immutable being is the abstract ground that allows phenomena to be saved: the nature always saved (*phûsis aeì sozoméne*; *Metaphysics*: 983b13), which is itself not hypothesised (*archè anhupóthetos*; *Republic*: 510b). According to Severino, the history of metaphysics in the West coincides with the positing of the different immutables of our tradition:

> In addition to immutable "truth" itself, the immutables through which the tradition of the West has attempted to defend itself from becoming include God, the immortal soul of man, the laws of nature, the laws of society regarded as natural laws, the laws of the unfolding of history, political and religious faiths, "common sense". Every immutable anticipates and predicts the meaning of everything that will gather around it (Severino 1988, 57–58).

As already remarked, however, the ground that affords to save the phenomena constitutes, at the same time, a manifest contradiction: firstly, because it is a being that is withdrawn from appearing and from becoming – and yet, *something* of it does appear, and therefore becomes; and secondly because, by virtue of its timeless being, it would stand in relation to everything that ever was, is or will be (i.e. every being would be related to the immutable itself). According to Severino, however,

---

7 "Being is forever kept and sheltered from the assault of nothingness" (Severino 2016, 46). In this framework, becoming itself consists in the appearing and disappearing of these eternal beings – the very appearing of beings being itself an eternal being that does not need to be saved. (It is in fact this very configuration of becoming that was deemed, by Severino's own mentor, Gustavo Bontadini, to be unable to save the phenomena in his "Sózein tà Phainómena"; Bontadini 1971.) See also Bontadini and Severino 2017.

8 We refer here in particular to the history of physics, but the same may be argued of the history of the other abstract enquiries or sciences.

the most certain form of self-evidence of the West is becoming itself: the arising of being from nothingness and their returning to it. If there exists an immutable being, the very nothingness from which beings are believed to emerge, and into which they are believed to disappear, would not be such: it would not be nothing. Even nothingness itself, by being related to the immutable, would come to have a content – it would not be the same nothingness that is unrelated to the immutable (in Severino's jargon, it would be 'entified' [*entificato*]). Once again, however, as part of becoming, beings are believed to arise from pure nothingness; if nothingness is entified through its relation with the immutable being, the most certain form of self-evidence of becoming turns into an impossibility:

> By anticipating the Meaning of the Whole, the immutable determines the Meaning to which the nothingness from which things originate as part of becoming must conform. The immutable thus transforms nothingness into a heeding and observance of Being: that is to say, it transforms nothingness into a being. What issues from nothingness, however, may not be compelled to submit to the Meaning of what already exists: for, otherwise, what arises would not be issuing from nothingness – i.e. from an absolute absence of meaning – but from the Meaning of what already exists. [. . .] If there exists an immutable, there may not exist any form of issuing from nothingness: that is to say, there may exist no becoming as it has been conceived once and for all by Greek philosophy (Severino 1992, 114–115).[9]

The very will of human beings, which believes to be able to produce every being (i.e. to bring it into being, according to the Greek meaning of *poíesis*: to pro-duce, *her-stellen*, etc.) finds an insurmountable limit to its power. It is for this reason, according to Severino, that the culture and civilisation of the West – having realised that the immutable and metaphysical dimension that has been posited beyond the manifest dimension of becoming and appearing turns the very appearing of becoming, and of the power and becoming of the human will, into a manifest impossibility – proceed to destroy that very metaphysical dimension. The last two centuries have borne witness to the destruction of every immutable dimension that would limit the power of becoming and of the will.

(II.) The second stage of the history of philosophy (and, consequently, of the different abstract sciences) coincides precisely with the endeavour to remove every immutable obstacle that would hinder the becoming and the power of the will. According to Severino, the apparatus of science and technics consists precisely in the practical and theoretical realisation of the conditions for an indefinite increase in the power of the will: an increase that may take place only insofar as every obstacle (i.e. every immutable being that would exceed the dimension of becoming) has been removed. According to Severino, the theoretical self-understanding of the sci-

9 A translation of *Oltre il linguaggio* into English is forthcoming.

entifico-technological apparatus is consistent with the self-evidence of becoming. That is to say that, while the existence of an immutable being pre-dicts and fore-sees the totality of being, thus entifying nothingness and giving rise to its own essential self-contradiction, the practical and theoretical procedures of the apparatus of technics aim in the first place to respect the structure of becoming (which entails complying with the very nothingness of nothing itself). Accordingly, the Apparatus (namely, the unified effort of science, technics and philosophy) functions according to the assumption that there exists no immutable being that would predict and foresee becoming itself. As such, becoming coincides with pure chance, "a falling upon existence without having been thrown by anything" (Severino 2023, 14), "a be-falling that falls out of nothingness" (Severino 2023, 47). The laws of this becoming may no longer be laws that determine what is and will be, but only laws of chance: for these laws "adequately express every regularity that can be ascribed to events that issue from nothingness: namely, the regularity described by statistical observations – the regularity of the taking place of chance as described by the laws of chance" (Severino 2023, 50–51). (Accordingly, physics, followed by the other abstract sciences, has had to turn its predictive apparatus into a statistical-probabilistic one.)

2 Saving the First Phenomenon

It should be remarked that the project of saving the phenomena is not a lofty project that starts to emerge in some of the most advanced human societies, and that is pursued only by astronomers and philosophers. The project of saving the phenomena constitutes, in fact, the very being (the very being-human) of humans themselves. That is to say, humans are such only insofar as they save phenomena: not insofar as they save this or that particular phenomenon, but insofar as they save the very *appearing* (the *phaínesthai*) of phenomena. Saving the appearing of phenomena tout-court is equivalent to verifying that *something* (or, in fact, *anything*) appears – that is to say, it is necessary to verify *that* phenomena appear. This verification constitutes the originary 'experiment' – namely, the originary *experience* (experiment, as well as experience, from the Greek *peîra*) – whose success is required in order to perform any other experiment or have any other experience. There exists an experience of phenomena only insofar as there is an experience of experience itself; some specific content appears only insofar as it appears that the appearing of phenomena appears. If it is not possible to verify that *something* (i.e. *anything*) appears, it is then also impossible to save any determinate phenomenon.

The originary experiment, or experience, thus posits the hypothesis that something appears, and proceeds to verify this hypothesis in order to save the phenomenon that consists in the very appearing of phenomena. This verification – namely, the experience of experience – underlies every other experiment, experience, or appearing of any phenomenon. It is only insofar as it is possible to save the appearing of phenomena (i.e. to verify that 'something', rather than 'nothing', appears) that it is possible to proceed to save any particular phenomenon.[10] Humans have an experience of different contents or phenomena insofar as they are first of all convinced that *something* appears to them – i.e. insofar as they believe to be constantly verifying the primary experience of experience. That is to say, however, that it is only insofar as they *fail* to save that first phenomenon, and *fail* to see that failure, that they may believe to proceed to save other phenomena (and fail to do so). Believing to have saved that first phenomenon constitutes the very essence of human beings.

In order to save the appearing of phenomena, in the most general terms, it is necessary to verify that *something* (i.e. *anything*) appears: 'something appears'; and, again, one is to verify that 'something (still) appears'. That is to say, one is to verify that, even if nothing has remained the same, *this* appears; it appears that nothing has remained the same. "Nothing has remained the same": this is the minimal hypothesis that underlies every experience. Even if nothing has remained the same, *this* must nevertheless appear. The hypothesis underlying this experience must be, on the one hand, that the *totality* of being has not been annihilated (i.e. that *something* has persisted into existence), and, on the other hand, that the totality *of its appearing* (i.e. of the 'one', the 'I' that experiences this experience) is also not altogether annihilated, but rather must, at least in some respect, have persisted into existence. If either the totality of what appears or the totality of the 'site' to which that content appears were to be annihilated – i.e. if nothing of what appears or nothing of the 'site' to which appearing appears were to continue appearing – the originary experiment or experience would not be verified. There must be a substratum that always persists both in the content or matter of appearing as well as in the form or appearing of appearing. The originary hypothesis thus presupposes – for a hypo-thesis is, precisely, a pre-sub-position – both a substance of experience that is preserved *and* a subject to which that experience appears. Sub-stance and sub-ject are the 'two' hypotheses of the originary experiment: the sub-strates (*hupokeímena*) that must be presupposed to any experience.

---

10 Accordingly, the fundamental question of metaphysics appears to enquire into the ground of the appearing or being of something rather than nothing (Leibniz).

This may also be stated in the following way: the original phenomenon is saved insofar as 'appearing' and the 'appearing of appearing', the transcendental matter and the transcendental form of appearing (the transcendental object and the transcendental subject of experience), are hypo-thesised and pre-supposed (that is to say, they are *abstracted*, i.e. extracted or saved, from the concreteness of experience). The abstraction of appearing from the appearing of appearing (the abstraction of being from its own being) thus constitutes the originary abstraction: the originary hypothesis that must be presupposed in order to save the appearing of phenomena. Given this hypothesis – namely that the substance of appearing is preserved, and that it appears to the same formal unity (i.e. to the same 'subject' or 'I') – the appearing of appearing, or the experience of experience, is saved (even if completely indeterminately and with regard to no specific content).

3 Beyond Saving the Phenomena

The originary abstraction, however, constitutes the originary contradiction. Every abstraction – every abstract hypothesis that is presupposed to the concreteness of appearing – is self-contradictory insofar as something 'is' (something is claimed to 'be') *even when it does not appear*.[11] The abstract notions of appearing and of the appearing of appearing (i.e. of a transcendental object and of a transcendental subject) are two self-contradictory notions precisely insofar as they are forms – and, in fact, the originary forms – of the contradiction of an hypothesis that attempts to save the phenomena.

This contradiction may not appear: it may not appear that something is abstracted from the concreteness of appearing. The concrete appearing of phenomena – the concrete appearing of appearing – does not need to be saved (and, in fact, it may *not* be saved) by abstracting or presupposing a certain hypothesis (i.e. by presupposing something to the concreteness of appearing; by hypostatising something that persists through time). The concrete appearing of appearing may only be saved by its immediate and timeless self-coincidence. According to Severino, this immediate self-coincidence and self-being of every appearing being constitutes its eternity: namely, the impossibility that *any* being may not be. Every being immediately coincides with the very ground that saves it; every being immediately coincides with its self-being; every appearing content immediately co-

11 In the same way as e.g. the thing in-itself lies beyond the appearing of the phenomenon – and yet, at least its doing so must appear.

incides with its self-appearing. This self-coincidence, through which every being immediately saves itself, is its eternity.

To be eternal is therefore the predicate of every being that is; to be, according to Severino, means to be eternal. The histories of physics and metaphysics are thus the histories of the failed attempt at abstractly saving the appearing of phenomena. As already remarked, there is one singular turning point as part of these histories, which consists in the realisation that phenomena may *not* be saved (a realisation that takes place with the closure of the histories of classical physics and traditional metaphysics). Contemporary science and philosophy recognise that phenomena may not be saved, and yet, according to Severino, they fail to recognise that every being and phenomenon is concretely and immediately saved by its own being, and they carry on positing the impossibility of saving the phenomena as the only abstract certainty that appears. Accordingly, to the eye of contemporary science and philosophy, everything appears in its own essential ungrounded-ness, as stretching out of nothingness and returning to it by pure chance (for every persistence would be a phenomenon whose appearing would need to be saved). According to Severino, this is the only picture of the world that is consistent with the one certainty that humans may not relinquish: the certainty of (or, in fact, the faith in) the existence and self-evidence of becoming – the only phenomenon that humans must continue to save, for otherwise they would be unable to save the very essence of their being-humans. It is only insofar as every being is abstracted from its being – only insofar as appearing is abstracted from the appearing of appearing – that it may appear (or, in fact, that it may appear to appear) that a being may be other than what it is: that a being may emerge from nothingness and return to it. The originary abstraction constitutes the very essence of human beings: the essential contradiction that human beings may not relinquish (and that, on the contrary, they try to eliminate by extending it to the totality of being, thus aiming to be consistent with their own ungrounded-ness). And yet, according to Severino, this abstraction or contradiction – the abstraction of every being from its being – may only be negated by the immediate self-identity of every being: i.e. by the concrete impossibility for every being to not be; by every being's having always been saved by its eternity. The concrete appearing of this eternity, however, may therefore appear only if the very being-human of human beings, the very contradiction that appears to save the being of human beings, ceases to appear: only if the history of the being-human of human beings comes to a close.

References

Bontadini, Gustavo. 1971. "Sózein tà Phainómena". In *Conversazioni di metafisica*, Vol. 2, 136–166. Milan: Vita e Pensiero.

Bontadini, Gustavo, and Emanuele Severino. 2017. *L'essere e l'apparire: Una disputa*. Brescia: Morcelliana.

Duhem, Pierre. 1969. *To Save the Phenomena: An Essay on the Idea of Physical Theory from Plato to Galileo*. Chicago: The University of Chicago Press.

Severino, Emanuele. 1981. *La struttura originaria*. Milan: Adelphi.

Severino, Emanuele. 1988. *La tendenza fondamentale del nostro tempo*. Milan: Adelphi.

Severino, Emanuele. 1992. *Oltre il linguaggio*. Milan: Adelphi.

Severino, Emanuele. 2016. *The Essence of Nihilism*. Edited by Alessandro Carrera and Ines Testoni. Translated by Giacomo Donis. London: Verso.

Severino, Emanuele. 2023. *Law and Chance*. Edited by Giulio Goggi, Damiano Sacco, and Ines Testoni. Translated by Damiano Sacco. London: Bloomsbury.

Simplicius. 2004. *On Aristotle On the Heavens 2.1–9*. Translated by Ian Mueller. London: Bloomsbury.

Simplicius. 2009a. *On Aristotle On the Heavens 3.1–7*. Translated by Ian Mueller. London: Bloomsbury.

Simplicius. 2009b. *On Aristotle On the Heavens 3.7–4.6*. Translated by Ian Mueller. London: Bloomsbury.

Simplicius. 2011. *On Aristotle On the Heavens 1.2–3*. Translated by Ian Mueller. London: Bloomsbury.

Leonardo Messinese

The Absolute Appearing of Eternity as the Original Meaning of Time

Abstract: The author of the contribution initially reflects on the distinction between the scientific knowledge and the philosophical-metaphysical knowledge of "time". He dwells on the trait that unites the Platonic and Aristotelian conceptions of time and, then, on the critical analysis that was done by Henry Bergson. The next step consists in comparing the Bergsonian reflection on time with that contained in Emanuele Severino, in order to introduce a "metaphysical" thought of time in which it becomes evident that the latter possesses its "original meaning" in the Eternal.

1 Metaphysics and Science of Time

A philosophical-metaphysical treatment of time implies the consideration of these three elements, which constitute its essential structure: 1) that something like "time" *appears*, manifests itself, bearing in mind that the experimental and theoretical results of science contribute to increasing the content of appearing, that is to say of the "unity of experience"; 2) that this content of appearing is thought of in its determined relationship with *Being* (*esse*); 3) avoiding that, in this unitary consideration of the two spheres of the "original structure" of metaphysical knowledge, we come to cancel – even if only implicitly – time in its phenomenological dimension.

For metaphysical thought, therefore, it is not sufficient to refer to the *phenomenological datum* of the temporal "development" for the purpose of an adequate understanding of time, but it is also necessary to establish its *ontological status*. In fact, in its specific way of referring to what is contained in appearing, it considers it "as being" (*ens*), that is, as regards its Being; moreover, we must immediately specify that the "meaning" of being that concerns metaphysical thought must not be presupposed on the basis of a different horizon of "meaning of meaning". In this way, it is possible to make an original distinction between a scientific treatment of the phenomenon of time and a philosophical-metaphysical one, since they respond to investigative perspectives that are formally different.

For scientific investigation, understanding time means building a "conceptual model" that allows us to unify a series of phenomena, which can be modified when it no longer responds adequately to this function. It therefore belongs to

https://doi.org/10.1515/9783111313610-004

scientific investigation as such that there is an *evolution* regarding the "concept of time", which depends both on a different way of "theorizing" about known phenomena, and on an "increase" of the phenomenal content that underlies the constitution of scientific conceptuality (Radicati di Brozolo 2001, 120–130). For the philosophical-metaphysical investigation, on the other hand, understanding time means asking oneself – as, for example, Aristotle did – not only what *defines* that determined content of appearing that we call "time", but also and above all it means asking oneself whether time belongs to the realm of Being and how it belongs to it (Aristotle 1984, 816); and therefore it means to establish the conceptual structure that allows us to understand both the Being that is "in" time, and the Being "of" time. Consequently – even if the theme cannot be explored here – what turns out to be decisive in the philosophical-metaphysical investigation, is ultimately what is the concept of "Being" that functions in it and presides over it.

Finally, in the context of these preliminary indications we must at least mention the fact that, on the third element of the philosophical investigation that I have indicated above, the accent is placed above all by *contemporary* philosophical thought, which even when it also articulates an "ontology", believes that in order to effectively protect the phenomenological dimension of time, the form of "theorizing" of temporality which is typical of metaphysics and, above all, of that contained in the classical tradition, must be abandoned.

2 Plato, Aristotle, Bergson

The phenomenological investigation of time, both when it is operated within metaphysical thought and when it emancipates it, can refer to *specifically different* contents of appearing. The diversity consists, first of all, in the fact that phenomenological inspection can refer to the "time of the cosmos", which is universal time; or to the "time of consciousness", which is lived time. In the economy of my writing, however, it will not be necessary to dwell on this distinction in relation to the *content* of the phenomenological dimension to be privileged; nor on Paul Ricoeur's attempt to overcome what he considers the aporia implied by the irrepressibility of the aforementioned distinction of perspectives on time, through the mediation of the "story" (Ricoeur 1988). My intent, in fact, is to thematize the Being of time starting from a highlighting of the *formal dimension* of metaphysical knowledge.

When Plato affirmed, in the *Timaeus*, that time is "a moving image of eternity according to number" (Plato 1888, 119),[1] in giving such a definition of time, he attributed the latter, as the content of appearing, to the Being of the cosmos. In turn Aristotle, when in Book IV of *Physics* he defined time as "number of motion [that is, of becoming] in respect of 'before' and 'after'" (Aristotle 1984, 821), he was indicating, albeit introducing a specific difference regarding the relationship between becoming and time, the same "phenomenon" that Plato had looked at – the time of *Physis* (Aristotle 1984, 827). Let us now consider a contemporary philosopher like Henri Bergson. He believes that the time of which Plato and Aristotle speak, is in reality the spatialized and mathematized time and not the time as an *original phenomenon*, which he sees in "duration".

I come now to the second question. Time as the content of appearing in Platonic thought was placed in relation to the immutable and eternal being of Ideas, as its "moving image". Going *beyond* Parmenides, but not *against* Parmenides, in this relationship Plato sees the speculative condition for which the affirmation of time no longer constitutes a denial of the truth of Being. Aristotle, on his part, would have proceeded in a similar way. In fact, the definition of the nature of time – of being "linked to movement", but also of distinguishing oneself from it, inasmuch as it is precisely its "measure" (Aristotle 1984, 818, 827) – constitutes only the first moment of the Aristotelian treatment of time. The Stagirite does not even on that aspect fundamentally departing from Plato, in Book VIII of *Physics*, in order to ensure the permanence of the totality of time and movement, of which time is numbering (and, therefore, to preserve the permanence of the universe), places them in the necessary relationship with the first immobile Mover: that is, Aristotle places time and movement with what constitutes for him the dimension of immutable Being (Aristotle 1984, 949, 950–952). The main difference between the two philosophers is that, while for Plato time is generated (Plato 1888, 121, 123), for Aristotle it is eternal, like becoming (Aristotle 1984, 922–923).

In this synthetic exposition, the role played by the soul in the Aristotelian consideration of time could not emerge (Aristotle 1984, 832–833). However, I would like here to highlight, on the one hand, what is the "phenomenon" in relation to which the Greek conception of time is mainly characterized; and, on the other hand, the inseparable relationship present in it between the "phenomenological" dimension and the dimension of "speculation" in the treatment of time, which persists in subsequent philosophical thought even when the concrete determination of the relationship of time with eternity will come to undergo variations.

---

1 Unless it's stated otherwise, all translations are mine.

Henry Bergson believed that, in the tracing of time to "immutable Being", the phenomenological dimension of time is, ultimately, totally annulled. This also means that Bergson, while criticizing the Platonic and Aristotelian conception, had grasped its underlying meaning very well. At the same time, however, in his firm intention of "saving phenomena" as a metaphysician himself, Bergson intended to establish a certain modality of the relationship between "time" and "Being", although of a different character from that of classical metaphysics, but then also of the modern one.

3 Bergson's Critique of Ancient Metaphysics

These initial reflections allow us to clarify the meaning of the title of my essay and of the speculative context in which it comes to be configured. Despite the basic objection raised by the major course of contemporary philosophy towards classical metaphysical thought, the essential thesis I intend to propose is that the Eternal, the immutable being, constitutes the original sense of time.

In order to carry out this thesis, it is necessary to first highlight the more precise meaning of contemporary criticism which sees metaphysical thought, inaugurated by Plato in substantial continuity with the conception of Being advanced by Parmenides, constitutively incapable of "saving the phenomena", as it would claim with respect to the thought of Eleate himself. Starting from the Parmenidean conception of immutable Being as "true Being", the metaphysics of Plato, but also of Aristotle – limiting ourselves, for now, to Greek philosophy – would have conceived in a prejudicial way erroneously the Being of phenomena, of the manifestation of the world, or ultimately the Being of becoming and, therefore, the Being of time. Compared to true Being, which is the immutable, or the eternal, the world of becoming would constitute degraded Being, which would add nothing to Being already accomplished in itself.

We can read in this regard a first text by Bergson, contained in his most famous work, entitled *Creative Evolution* (Bergson 1944). The philosopher, in proposing a new metaphysics, is the voice of contemporary criticism of the conception of Being as immutable. One of the ways in which the underlying sense of Platonism is indicated by Bergson is the following:

> To reduce things to Ideas is therefore to resolve becoming into its principal moments, each of these being, moreover, by the hypothesis, screened from the laws of time and, as it were, plucked out of eternity. That is to say that we end in the philosophy of Ideas when we apply the cinematographical mechanism of the intellect to the analysis of the real (Bergson 1944, 342).

This *resolution* of things into ideas, of phenomena into forms, of becoming into immutable Being, of time into eternity entails, according to Bergson, nothing less than the very *fading away* of things, phenomena, becoming, time. The philosophy of "ideas", covering things with its own view, is unable to restore them to us, but only to conceal them. Instead, Bergson writes, "life is an evolution" (Bergson 1944, 328).

The error of metaphysical thought would therefore be analogous to the error made by the common conscience. It would consist in concentrating "a period of this evolution in a stable view which we call a form, and, when the change has become considerable enough to overcome the fortunate inertia of our perception, we say that the body has changed its form" (Bergson 1944, 328). In reality, for Bergson, it should not even be said that it is constantly, in every instant, that "the body changes shape", but rather that "there is no shape". The form, in fact, is immobile, while "the reality is movement" (Bergson 1944, 328), the reality is the process of becoming, it is *changement* (Bergson 1946, 153–185). Here, we can legitimately add: if, for Bergson, the form is to the immutable, as the becoming is to the time, then for him *Being is time*. It is not time that must be resolved in the immutable, in the eternal Being, but it is Being that must be thought of *as* time.

This last expression, in Bergson's perspective, means that thought does not hide the *fact*, but is its authentic manifestation. It means that thought does not impose on "phenomena" the "categories" of thought, but lets them be as they are. In other words, it means that theory does not replace things, but rather is their *manifestation*. And what, then, is the "reason"? To such a question, Bergson replies as follows:

> The reason is that there is *more* in the transition than the series of states, that is to say, the possible cuts [practiced by intelligence within the continuous flow of becoming, AN] – *more* in the movement than the series of positions, that is to say, the possible stops [equally operated by intelligence, AN] (Bergson 1944, 340–341).

This "more", this irreducible *excess* that opposes the "resolution" of phenomena in the immutable that is operated by metaphysics, is none other than "time": time by Parmenides immediately and absolutely denied; and time also, albeit, in a more tenuous but no less effective way, denied by Plato and Aristotle. Within metaphysical thought – Bergson notes – extension and relaxation, space and time, "simply manifest the gap between what *is* and what *should be*": between degraded Being and Being in its perfection. But the presumed "metaphysical" salvation of degraded Being of what is in space and time, the presumed "metaphysical" salvation of the world of becoming, turns out to be, instead, for Bergson the irredeemable loss of it.

Such a thought, that carries out that progressive work of cutting off the unitary flow of becoming and time, is constituted in its canonically metaphysical form when it reaches, on the one hand, to the "system of ideas, logically co-order together or concentrated into one only" (Bergson 1944, 355); and, on the other, to "a quasi-naught, the Platonic 'non-being' or the Aristotelian 'matter'" (Bergson 1944, 355). Within metaphysical thought this vivisection will be followed by an attempt to recompose the initial unity, through the structuring of a system in which the series of degrees into which the immutable degrades (Bergson 1944, 356) is placed between the Being of the principle and the non-Being. Such a cosmology – Bergson equally writes – will always be dominated by the immutable principle and the "physical" will always be defined by the "logical" (Bergson 1944, 356). This is why time, in Greek metaphysics, is considered "theoretically negligible". For the Greek philosophers, as has been said, "the duration of a thing only manifests the *degradation* of its essence" (Bergson 1944, 373).

4 Bergson's Critique of Modern Metaphysics

If we go to inspect Bergson's position with regard to modern metaphysics, it is noted in the first place that, for the French philosopher, in its concrete articulation it did not depart from Greek metaphysics in its essential form; this, although due to the advent of Galilean science it had found itself in a position to work a new path.

To adequately understand this assumption, it is a question of seeing what constitutes the most innovative feature of modern science with respect to the ancient form of knowledge. Of course, compared to the knowledge of the ancients, modern science does not concern the "concepts" by which to know things, but it seeks "laws", understood as the constant relations between variable quantities. For example, for Aristotle "the concept of circularity was enough to define the motion of the stars", while for Kepler, to establish the "movement of planets", it was not enough to have the *concept* of "elliptical form" (Bergson 1944, 362). But the major difference, for Bergson, does not consist in this; or, better said, all this is merely the consequence of something else: it is the consequence of having made the *"time" factor* intervene in the calculation of the relationships between the magnitudes of the various object fields, while the science of the ancients is, on the contrary, a *static* science (Bergson 1944, 363). The time factor intervenes in Kepler's astronomy, in Galileo's mechanics and, in a more veiled way, in Descartes' geometry itself (Bergson 1944, 363).

Immediately after showing the difference between modern science and ancient science, however, Bergson asks himself: "But with *what* time has it to do?" (Bergson 1944, 365). Well, the time of physics is not the time of "duration", that is, of continuous creation, of creative evolution: in other words, it is not what Bergson calls "time-invention". The time of physics is that with which we find ourselves once we abstractly consider the events isolating them from each other, that make up the whole, considering them "such as they would be outside of the living whole, that is to say, in a time unrolled in space" (Bergson 1944, 371). This means, for Bergson, that modern physics *"rests altogether on a substitution of time-length for time-invention"* (Bergson 1944, 371). Faced with the time of physics, "You can say that you are still dealing with time [. . .] but we shall know that we are no longer dealing with an experienced time; we shall be before a symbolic and conventional time, an auxiliary magnitude introduced with a view to calculating real magnitudes" (Bergson 1965, 64).

This spatialized time of physics, the time that flows uniformly from the past to the future, theorized by Newton as "absolute", independent of things, in short, the time measured by clocks, has the consequence for Bergson that, if becoming does not exist, the past, the present and the future would be *present together*, in a similar way to the frames of a cinematographic film. On the contrary, for the "time-invention" the future is not fully determined, it is open to the unpredictable, it does not simply consist of what is attributable to pre-existing elements.

5 Bergson and Severino on Time and Metaphysics

By bringing to extreme clarification what is contained in the Bergsonian concept of time-invention, Emanuele Severino will come to note in the position of the French philosopher a convergence with the "Greek sense of becoming". This is because, beyond his negation of Greek metaphysics, which sees in the "eternal" the *ultimate meaning* of what becomes, of the things that are in time, also by Bergson the becoming is "understood as the creation of the unpredictable nothing and as annihilation" (Severino 1996, 806). For this reason – Severino adds – "Bergson's controversy against the idea of absolute nothingness does not mean that for him becoming is not annihilation" (Severino 1996, 806). So that, for Severino, the ultimate root of the thesis of the unpredictability of the future, which scientific knowledge ultimately reached by affirming the statistical-probabilistic status of its laws, lies in the conception, which was also affirmed by Bergson, that the future is literally *nothing* (Severino 2016d, 41; Severino 2023).

The Severinian thesis of an irreducibility of the "nothing" in Bergson, which remains affirmed *in actu exercito* at the very moment in which he denies it *in actu signato* polemicizing with the thought of ancient metaphysics – and which implied it in correspondence with the immutable Being, to justify the degradation of the latter into multiplicity and becoming – is, in my opinion, convincing. In fact, if the *mouvant* – the becoming – for Bergson "can only *transit* in the inexhaustible multiplicity of stations or figures that its transit *brings into Being*" (Ronchi 2008, 55) how to escape the implication of nothingness to ensure this "bringing into Being" of multiplicity and of becoming? Unless creation is not understood at all in Bergson's way – that is, as "creative evolution" – but in a different way, which I will focus on towards the end of my paper.

For Bergson, moreover, modern metaphysics repeats the same error of ancient metaphysics through different concepts, that of canceling the Being of time into eternal Being. For both metaphysics, "*reality as well as truth are integrally given in eternity*. Both are opposed to the idea of a reality that creates itself gradually, that is, at bottom, to an absolute duration" (Bergson 1944, 384–385). The Bergsonian "duration", on the other hand, expresses the very meaning of becoming as well as of time. As for the first, in Severino's exposition of it, "becoming is not the juxtaposition of different and separate states and instants, but it is a *continuous flow* where the past is projected into the present and penetrates it" (Severino 1996, 801). As regards time, Bergson argues that "'duration is the continuous progress of the past that gnaws at the future and enlarges it as it advances'" (Severino 1996, 801). The sense of "continuous creation", in Bergson, is "the continuous realization and appearance of what did not exist and which is precisely for this reason new and unpredictable" (Severino 1996, 801).

The assumption of the centrality of time leads Bergson to design a new metaphysics. Perhaps, however, in formulating his own philosophical doctrine, Bergson did not draw all the consequences that could be drawn from the escape of the Being of time from the orbit of "metaphysical Being", that is to say, from the assumption of "seriousness" of time, which acts as a counterpart to the seriousness of becoming as it was understood within the ontological thought of the Greek philosophers (Severino 2009b, 151–166).

Deepening the question of the way in which the relationship between time and the eternal has been analyzed throughout the entire history of Western thought, Emanuele Severino for his part would have shown that the "seriousness" of becoming and of time entailed not only the end of metaphysics in the form inaugurated by the Greeks, but also the end of every hybrid mixture of time and eternity that characterizes every form of metaphysics. Bergson, on the other hand, had held that the emancipation of time from the eternal finally entailed the establishment of true metaphysics, distinct from scientific knowledge and, even more so, irreducible to the latter, thus restoring a primacy of philosophical speculation with regards to scientific knowledge.

When Severino refers to Bergson, he emphasizes above all his belonging to the most consistent current of contemporary philosophy, for which the force of the (supposed) *experience* of becoming nihilistic – that is the (supposed) *original evidence* of "time" – has won over the forms of traditional knowledge which had claimed to enclose, with the "net" of their immutable determinations, the "sea" of becoming. I believe, however, that it can equally be argued that, in Severino's eyes, the Bergsonian concept of duration also constitutes a contradictory mixture of time and eternity; this, for that remnant of metaphysical "theory" that concept of duration contains, in contrast with the thesis of the seriousness of becoming and of time based on the non-anticipability of the future. This is based in turn on the conviction – for Severino of a nihilistic character – of radical emergence of phenomena from *nothing* of themselves.

6 Severino's Critique of Methaphysical Thinking

The next step in this paper, in a consequential way, will be to take a look at the Severinian conception of time, immediately specifying that, what is stated by Severino when referring to time, more precisely concerns *the "Greek sense" of time*. This meaning, for the Brescian philosopher, is contained in the ontology ideally inaugurated by Plato in the *Republic* (Severino 2016c, 149–206) and reaffirmed by Aristotle in his own formulation of the "principle of non-contradiction" (Severino 2016a, 35–83). This same sense has been preserved within Western philosophical thought and culture, even when we have moved away from the metaphysical superstructure that, in Greek philosophers, encompassed the region of temporal being. Moreover, for Severino it must be affirmed that, even when man does not yet explicitly express the alienated sense of time – as it happened before the advent of Greek ontology – or he introduces some conceptual structures which, in any case, integrate the original sense of becoming, he is *in any case* an "inhabitant of time". Let's look at the Severinian thesis a little more closely, approaching a short paper entitled *Time and alienation* (Severino 2009a, 31–44).

Taking into consideration Aristotle's *De interpretatione*, Severino points out that for the Aristotelian text there is no problem with a non-nothing, for example the house, "when it is not, it is not" (Severino, 2009a, 31), that is, it is not absurd that a *non-nothing*, it need not be. And, however, the house, the stars, men, the earth are *beings* and, as such, they are united with *Being*. Therefore, Severino wrote in a lapidary way: "*Tempus* separates beings from their Being" (Severino 2009a, 36); and again: "inhabiting time means separating being from Being" (Severino 2009a, 37). The man for whom time, understood in this way, constitutes the original evidence; the man for whom it is clear that being is "what goes out and returns into nothingness" and fails to see the absurdity contained in this belief, is the "inhabitant of time" (Severino 2009a, 38). And it is because of this original separation of being from Being of which man is convinced, first implicitly and then explicitly, that he then "needs God (or revolutionary praxis, or technology)" (Severino 2009a, 40), that is, the forms by which man has tried to protect himself from nothing by planning, in different ways, to dominate the oscillation of beings between their Being and their non-Being.

All this, however, fails when the "testimony" of the truth of Being and, in very close connection, of the truth of being appears. Within this testimony, "every being is eternal [. . .] and the variation of the spectacle of the world, the appearing of the variation is the sunrise and the sunset, the showing and hiding of the eternal Sun" (Severino 2009a, 43). In the "Postscript" to "Returning to Parmenides" Severino offered for the first time the argued articulation of this thesis (Severino 2016b, 85–145). Here we can only recall his basic sense:

> When being comes out of the vault of Appearing, the Appearing is silent about the fate of the hidden being (and the 'when' acquires an unprecedented meaning). But the Erinyes of truth [. . .] of which Heraclitus speaks (fr. 94) reach what is hidden and remind it of his destiny: the need, the *Ananche* that it remains united to his Being (Severino 2009a, 44).

7 Time, Eternity and Creation

Both in Bergson's thought and in Severino's thought, there is a critique of the Greek conception of time. And, in both, the criticism is intimately linked to the way in which time is referred to by Plato and Aristotle to the dimension of the eternal. Nonetheless, what their respective criticisms draw from is profoundly different.

Bergson observed that, in the *tracing* of time to the eternal, of the becoming to the immutable, time and becoming are *reduced* to eternity, that is to say they are made null, they lose themselves: time and becoming are only being "degraded" with respect to "true" Being, the repetition to a lower degree of Being of what has always been. The *relationship* between time and the eternal, for Bergson, nullifies the Being of time.

Severino, for his part, also pointed out that the relationship between time and the eternal cannot exist, if becoming is understood as the beginning to be and the ceasing to be of the being and, consequently, time is what which separates being from its Being. The relationship, however, must fail, not because it is the eternal that renders time null, but because it is the affirmation of "time" – of what separates being from its Being – that renders nothing, that is, impossible, the affirmation of the eternal, of the immutable and to leave the field, without realizing the absurdity that is involved in it, only to being identified with its *nothing*: i.e. the non-nothing identified with the nothing. And this absurd identification is implicitly affirmed even when he intends to deny it by affirming the principle of non-contradiction, which opposes Being to not Being, in the manner of Aristotle.

In the presence of these two distinct outcomes that show the theoretical tangle in which metaphysical thought inaugurated by the Greeks is involved – outcomes that I call distinct in that Bergson starts from the evident statement of “time” and Severino, on the other hand, relies on the affirmation necessary of the “eternal” – it would seem that the very ground on which to base the thesis that constitutes the title of this essay is originally lacking: “The absolute appearing of the Eternity as the original meaning of time”. But is this really the case, or can a different conclusion emerge from all the events of the time that I have tried to tell? And this, perhaps by acknowledging the Severinian indications regarding the uncontradictory affirmation of the Being of entities, but then also the Bergsonian ones that insist on the reality of the time?

Introducing this question presupposes, first of all, the possibility of a refinement of metaphysics that adequately dissolves the theoretical knot of the relationship between the becoming and the immutable, or between the order of time and the dimension of the eternal. Such a possibility appears when the affirmation of the becoming being – of what is measured by time – is freed from an erroneous conception both with regard to “what is measured”, or the becoming being, and to “what is measure”, that is time. The becoming does not appear as that which has, at its two extremes, the not Being from which it comes and the not Being to which it would return. And time, in its turn, is not what isolates the becoming, at the two extremes, from the Being of it.

In this regard, I can only refer here to some of my writings in which such a non-nihilistic conception of phenomenological becoming has been justified and, in relation to it, the essential terms of the relation of time to eternity have been established, of the becoming to the immutable. It is introduced “speculatively”, that is to say, indicating the reason why what becomes, what belongs to the order of time, is necessarily united to the eternal: it is *in relation* with the eternal in the sense that it *is* related to the eternal (Messinese 2008, 307–329; Messinese 2009, 533–555; Messinese 2015, 141–152; Messinese 2019, 211–235). Here it is only possible to identify some essential points.

1. The *concept* that is speculatively introduced to affirm the relationship of the temporal to the eternal is itself *the* relationship in which being is incontestably affirmed in its Being.
2. The concept in question is introduced *speculatively* because, like any metaphysical concept, it does not belong to the field of “expressive thought”, that is, that which concerns the Being of experience, but to the field of “demonstrative thought”.

3. This speculative concept is that of *creation* – freed from an erroneous "fabricator" understanding of creating – which affirms the very "principle of Being of the being".

According to a profound formulation by Gustavo Bontadini:

> The Principle of Creation is the fulfillment of what must be considered as the Principle of Parmenides in the proper, historical sense – Being is immutable, Being cannot fail to be – as it allows us to gain that perspective, it gives us access to that point, in which we see that this last principle is no longer falsified by experience (in which we see that the only reality is God: in which, that is, not only "the existence of God is demonstrated", but even that existence – the *actus essendi* – is gathered entirely in God, that "God is the Existent") (Bontadini 1995, 194).

Creation, therefore, must not be understood in analogy with what is called the "producing" of nature or human action – even assuming that such a production must be assumed according to the common way – and, that is, it must not be understood anticipating that meaning to the properly "metaphysical" meaning of the term. Creation, on the other hand, must be understood according to the meaning that appears in the act in which its concept is introduced speculatively due to the non-contradictory nature of Being.

When this settlement of classical metaphysical thought is achieved, which implies the theoretical adjustment of Parmenides' Principle itself, the nihilistic meaning of time is also lost, which constituted for Severino the root of the impossibility of uniting time with eternity.

References

Aristotle. 1984. "Physics". In *The Complete Works of Aristotle. The Revised Oxford Translation*, edited by Jonathan Barnes, 699–982. Princeton: Princeton University Press.

Bergson, Henri. 1944. *Creative Evolution*. Translated by Arthur Mitchell, with a Forward by Irwin Erdman. New York: Random House.

Bergson, Henri. 1946. "The Perception of Change". In *The Creative Mind: An Introduction to Metaphysics*, translated by Mabelle L. Andison, 153–185. New York: The Philosophical Library.

Bergson, Henri. 1965. *Duration and Simultaneity: Bergson and the Einsteinian Universe*. Translated by Leon Jacobson, with an introduction by Herbert Dingle. Indianapolis: Bobbs-Merrill.

Bontadini, Gustavo. 1995. *Conversazioni di metafisica*, Vol. 2. Milan: Vita e Pensiero.

Messinese, Leonardo. 2008. *L'apparire del mondo. Dialogo con Emanuele Severino sulla "struttura originaria" del sapere*. Milan: Mimesis.

Messinese, Leonardo. 2009. "La teologia razionale e la determinazione dell'Altro dall'esperienza". *Rivista di filosofia Neo-Scolastica* 101(4): 533–555. http://www.jstor.org/stable/43063946, last accessed 19 September 2023.

Messinese, Leonardo. 2015. *L'apparire di Dio. Per una metafisica teologica*. Pisa: Edizioni ETS.
Messinese, Leonardo. 2019. *La via della metafisica*. Pisa: Edizioni ETS.
Plato. 1888. "Timaeus". In *The Timaeus of Plato*, edited by Richard D. Archer-Hind. New York: McMillan.
Radicati di Brozolo, Luigi Arialdo. 2001. "L'evoluzione del concetto di tempo in fisica". In *Il tempo dell'uomo e il tempo di Dio: filosofie del tempo in una prospettiva interdisciplinare*, edited by Adriano Fabris, 120–130. Bari: Laterza.
Ricoeur, Paul. 1988. *Time and Narrative*. Translated by Kathleen Blamey and David Pellauer. Chicago: University of Chicago Press.
Ronchi, Rocco. 2008. "L'oro e gli spiccioli. Bergson, la teologia aristotelica e il principio di continuità". In *Dio, la vita e il nulla. L'evoluzione creatrice di Henri Bergson a cent'anni dalla pubblicazione: atti del colloquio internazionale*, edited by Giuseppina Strummiello, 1–19. Bari: Edizioni di Pagina.
Severino, Emanuele. 1996. *La filosofia dai Greci al nostro tempo*. Milan: Rizzoli.
Severino, Emanuele. 2009a. "Tempo e alienazione". In *Gli abitatori del tempo*, 31–44. Milan: BUR Rizzoli.
Severino Emanuele. 2009b. "Attualismo e serietà della storia". In *Gli abitatori del tempo*, 155–166. Milan: BUR Rizzoli.
Severino, Emanuele. 2016a. "Returning to Parmenides (Postscript)". In *Essence of Nihilism*, edited by Ines Testoni and Alessandro Carrera. Translated by Giacomo Donis, 85–145. London/New York: Verso.
Severino, Emanuele. 2016b. "Returning to Parmenides". In Id., *Essence of Nihilism*, edited by Ines Testoni and Alessandro Carrera. Translated by Giacomo Donis, 35–83. London/New York: Verso.
Severino, Emanuele. 2016c. "The Path of Day". In Id., *Essence of Nihilism*, edited by Ines Testoni and Alessandro Carrera. Translated by Giacomo Donis, 149–206. London/New York: Verso.
Severino, Emanuele. 2016d. *Storia, Gioia*. Milan: Adelphi.
Severino, Emanuele. 2023. *Law and Chance*. Translated by Damiano Sacco. London: Bloomsbury.

Massimo Cacciari

Note on the Dialogue between Severino and Vitiello

Abstract: The chapter considers a discussion that took place in 2018 between two prominent Italian philosophers: Emanuele Severino and Enzo Vitiello. The object of contention concerns the relationship between the finite and the infinite with respect to the appearance of the entity, which does not appear in the totality of its determinations-relationships. The question concerning the relationship between being and entity is considered by traversing the contributions of Aristotle, Leibniz and Dante, with a reference to the Gospel and the Upanishads.

Enzo Vitiello observes against Severino (their conversation is fundamental: "Dell'essere e del possibile ("On being and possibility") (Severino and Vitiello 2018):[1] if the being appeared in the totality of its determinations-relations, it would necessarily appear as including in itself precisely this totality, and therefore could no longer appear as the finite existent that it actually is. But what difference can really be made between the finite and the infinite? The being is, and is undeniably itself – yet it never appears *simplex et unum*. Every being always appears marked: the candle is lit or extinguished, on a table or on a chair, in the light of dawn or sunset, and so on *ad infinitum*. The Principle of Non-contradiction forbids the possibility of saying the candle is at the same time lit or extinguished – and one could push its value as far as to say that the lit candle is not the extinguished candle, that both are – one could, that is, remove the temporal limitation that Aristotle assigns to the Principle. Nevertheless, those different determinations subsist. No one will ever see the candle as pure candle. To every being compete infinite determinations – infinite and yet finite, since the candle will never compete with the determinations that compete with an elephant. It is a finite infinity, just as even numbers are infinite, and yet a class of numbers. Nor can such determinations be accidental; for if they "happen" to an being, this means that it contains them in itself in some way, otherwise they could never happen to it. Prius is certain that the being is, but in its being are also its manifold determinations. The manifold appears in one with the identity of the thing, and the identity of the thing appears only according to a finite horizon. The thing's name is plural (as is that of the Ego, which is its manifestations and is likewise the unrepresentable Self itself); its present identity is the product of

1 Unless otherwise stated, English translations are by the author.

https://doi.org/10.1515/9783111313610-005

all that has been, nothing im-mediated – but all that has been really gives itself, ek-sists in the thing's present being and is here-and-now present, constitutes a necessary dimension of the same identity.

Not only that. The appearance of the thing is always also, at one and the same time, the appearance of its relation to something else (how could I affirm its identity if I did not compare it to the other from itself?). To identity belongs, together with the multiple, being-in-relation. Relation with the other, and certainly not with nothingness. Severino rightly insists on clarifying the radical difference: other does not mean nothing. It makes no sense to affirm that nothingness is nothing other than the expression par excellence of the Other. Nothing is nothing and not Other. The being has in itself the relation to other – its relation to nothing is nothingness. The relation is the relation of the thing to the All that is, to the totality of the being. We must therefore posit in the "original structure" (Severino 1981) that A is A in the totality of its determinations and relations. And therefore conceive it as containing the All in itself. As we have already seen in Leibniz (and Vitiello refers to Leibniz): "the singular substance implies all its predicates [. . .] in its perfect notion it implies the whole universe and all things existing in it, past, present and future" (Severino and Vitiello 2018). In the point of the singularity of the being the universe contracts, "in the universe the point is ex-tended" (Vitiello 2008). Does affirming this imply contradicting the *aìsthesis*, falling into a sort of logical-intellectual hybris that would be purely indicative of the impotence of reason (for both Aristotle and Kant, reason that does not give reason for the *aìsthesis* itself is only a sign of its own ignorance or impotence)? No; to affirm this is to combat the error that *doxa* commits precisely in understanding the modes of sensible experience, that is, in its regarding the thing as separate-isolated from the Whole, from time to time according to abstract determinations, making abstraction from the Real Most Real: not being the entity other than the contracted Infinite itself.

Now, however, the infinite appearance of the totality of beings is not given. That is, it does not appear to us, it is not concretely *phainòmenon.* The mind intuits it, in the full sense of *noein,* as necessary, but cannot concretely experience it. Only God, or the absolute Monad or the Causa sui, comprehend, in the different ways we have seen, every being in its ultimate, radical *ratio essendi* and therefore also in the totality of the connections that connect it to the All. The being is all this, but all this it appears according to a finite horizon, according to finite determinations, and so it must be predicated. The human mind, we repeat, certainly understands the "destiny of necessity" that orders the All, but how could it see and preach in one the infinite modes of appearance? "I saw that in its depth far down is lying / Bound up with love together in one volume, / What through the universe in leaves is scattered; / Substance, and accident, and their operations, /All interfused together in such wise" (Alighieri, Paradise XXXIII, translated by Henry Wadsworth Longfellow 1867, 85–89). And yet Dante himself only senses it, at one point, in the "instans", he cannot speak of it; he feels he enjoys it, he feels Gioia, but his speech is too short and dim compared to his vision (we will return to the theme of the missing word). How is true Gioia ("Joy") possible, then? Dante's stamp is eschatological, understanding the term in the precise sense that here, *in hoc saeculo,* of Gioia (which he would call "beatitude") there can only be real traces and a certain promise (but gratia and by faith). Outside of a horizon analogous to Dante's, will it be permissible for us to state that a trace of Gioia is to understand the destiny of necessity, as also for Spinoza the soul arrives at the condition of "Laetitia" or "hilaritas" in intuiting every being included in its Cause? But such *laetitia* continues to imply in itself, as we have seen, the transience of the being, and, on the other hand, Dante's enjoyment presupposes the act of faith. The Gioia of which Severinus speaks opposes "radicitus" to both perspectives: for him, everyone sits at the right hand of the Father *ab origine.* Is the memory of John 15 recalled here? "These things I have spoken to you because he charà he emè, so that the Gioia that is mine, Gaudium meum, may also be in you and your Gioia may be full". The Word is welcomed with Gioia, the same that filled the Magi at the sight of the star. It is the Word that changes affliction into Gioia (John 16:20). What then? Is Gioia an attained condition? Why then do we err from it, continuing to isolate the appearance of the being from its eternal being? What ontologically explains this error-error? Why do we persist in the belief that we are mortal, thus continuing in the belief that we have the power to give death? It is answered by the fact that being appears according to finite, contradictory horizons – and the removal of the totality of contradictions is Gioia. Gioia is therefore already in the original structure. One could call this position eschatological only in the sense that the actuality of the Ultimate is affirmed there, the truth is now spoken, the same truth that we shall see at the Ultimate. The *èschaton* coincides with the origin. In other words: the possibility of Gioia is not a possible, it expresses the actual power of being. And yet the shadow of error necessarily follows it, and will follow it until the contradiction between the finite appearance of the being (the order of Chronos) and the eternity of the All is overcome. So would Gioia be worth as the transcending of all duality, of all obstacles to the knowledge of the identity of seeing and seen, knowing and known, light of the individual and light of the Whole – analogy with the Fourth State of which the Upanishadic *sophia* is an analogy? "Truly this great and uncreated atman, without old age, without death, devoid of fear, is brahman. Verily the brahman is happiness and becomes the brahman himself, who is happiness, he who thus knows" (The Bṛhadāraṇyaka Upaniṣad (with the Commentary of Śaṅkarācārya) 1950, IV, 4,25). As long as one sees something other than oneself there can be no Gioia, as long as one remains chained to the finite manifestations in their contradiction one does not give *sukha* (*succus*, *sykon* in Greek is the sweet fig), one does not "drink" at the source of the knowledge of immortality. And yet Roberto Calasso in *L'ardore* has recalled how close to *sukha* is the term for suffering, *soka* – and how, on the other hand, in the eschatological condition of identity between the Self and the Whole could the pain experienced to reach it, the experience lived "in this perpetual flow like a frog in a dried up well" (Maitry-upanisad, I, 4) be erased? Even the Johannine Gaudium is inseparable not only from agape (to remain in Gioia it is necessary to remain in the same bond of Love that binds the Son to the Father), but also from the Cross (the Gioia in the Cross Cristina Campo speaks of, in some ways so akin to the Gioia we can attain, according to Spinoza, by knowing that God is the cause of sadness itself, *Ethica* V, 18, scholium). And Michelstaedter would ask: but can this Gioia concern the individual? Can the ego rest in something absolute? How could the individual enjoy this Gioia without eliminating his own individuality, without freeing himself from the particular love that binds him to the Self? More and more concretely the content of the original structure appears – says Severino. But what does this "more and more" mean? Does it mean that in some form of becoming, little by little, the contradiction is transcended? Then, inhabiting the becoming would take on a positive meaning for truth, and it would be the truth of the original structure itself that would only be valid as a trace of Gioia; up to the *èschaton*, Gioia itself could, therefore, only be given analogically.

Is a phenomenology of consciousness conceivable that would make the experience of the Gioia present *ab aeterno* in the original structure ever more concrete? To the concretely absolute truth that consists in the removal of the totality of contradictions corresponds the truth, equally absolute, but not concretely absolute, says Severino, of the finite circles of destiny. If this is so, however, Gioia will once again assume an eschatological-apocalyptic timbre and the destiny of necessity will still leave room for a *nondum* that seems to contradict it. If there is a transition between these different dimensions of absolute truth, then, *in hoc saeculo*, Gioia is impossible. It would only be conceivable if absolute truth rationally understood and absolute truth also concretely experienced (concrete is that which con-creates and co-exists with and in the totality of the determinations of the being) could be understood as states both eternally com-present. Then, the contradiction between the two dimensions would still remain, but as resolved *ab origine*. A kind of *kàtharsis* preceding the tragic mythos, a catharsis ante factum – a medicine that ignores the unpredictability of evil, the power of *Tyche*. Nevertheless, within this limit, a trace of Gioia can undeniably be given, where one understands how undeniably the being expresses its own substantiality and its point-individual is ex-tended to the infinity of its determinations and relations, just as certainly the effects of its action are ex-tended to infinity.

References

Alighieri, Dante. 1867. *The Divine Comedy*. Translated by Henry Wadsworth Longfellow 1867. Boston: Ticknor and Fields.

Maitry-upanisad. 1862. *Maitry-upanisad*. Maitry-upanisad (with the Commentary of Ramathirta, translated by E. B. Cowell). Calcutta: C.B. Lewis/Baptist Press.

Severino, Emanuele. 1981. *La struttura originaria*. Milan: Adelphi.

Severino, Emanuele, and Vincenzo Vitiello. 2018. *Dell'essere e del possibile*. Milan: Mimesis.

The Bṛhadāraṇyaka Upaniṣad. 1950. *The Bṛhadāraṇyaka Upaniṣad* (with the Commentary of Śaṅkarācārya) by Swāmī Mādhavānanda. Mayavati: Advaita Ashrama.

Vitiello, Giuseppe. 2008. "Essere nel mondo: Io e il mio Doppio". *Atque. Materiali tra Filosofia e Psicoterapia*, Vol. 5. https://d1wqtxts1xzle7.cloudfront.net/41879196/Essere_nel_mondo_Io_e_il_mio_Doppio20160202-3844-ru6rgd-libre.pdf?1454403697=&response-content-disposition=inline%3B+filename%3DEssere_nel_mondo_Io_e_il_mio_Doppio.pdf&Expires=1696770820&Signature=Fl1VZVA~jaKDhvrUzDuNEHOzSTAi5hy5soPbPXih5tvoBx0gNao4hB-IdfNkjSlePNAwYrQ5n-W3Gez7rWF212VUyMdS4unPfXLsSNiFEWoiJoKU9WMeOgRDcvZ2EwMFU81Bm0OfNfzqOTnwdTO1A8YWR2r8lgkkoPxF~r6WtrHzKYPoZRqt8LzoUEto8iUuPYtgJ~FdgaWswR339M7ee4oJJADDCAb5lzx5ggFghRMr7fYrCNclwcLRmyj5ChpuLnAE1ysky~~A0MsCn3e11zMmda~I2aelhicPEkv1moDLCrx0MHKAJjEW8jkpHj0obBXhmuHUevYZ~qD-C4tuqA__&Key-Pair-Id=APKAJLOHF5GGSLRBV4ZA, last access 8 October 2023.

Roberto Tommasi

Time, Eternity, Freedom in Kierkegaard, Heidegger and Ricœur

Abstract: The contribution investigates the correlation between time, space and eternity in Kierkegaard, Heidegger and Ricœur. From the perspectives opened by the three thinkers emerges the aporetic oscillation between cosmological, existential and historical conceptions of space-time.

The current contribution offers some indicators aimed at illuminating the problem of the relationship between the eternal, time and freedom – a very complex and intricate issue that runs through the history of philosophy – starting with Søren Kierkegaard, Martin Heidegger and Paul Ricœur, chosen because they are able to open up some interesting and original philosophical and theological perspectives on the subject.

1 The *Paradox* of the Eternal in Time (Kierkegaard)

Søren Kierkegaard in the *Philosophical Fragments* clarifies the intertwining of time and eternity in relation to the question of human being's possibility of eternal salvation.

> Can a historical point of departure be given for an eternal consciousness; how can such a point of departure be of more than historical interest; can an eternal happiness be built on historical knowledge? (Kierkegaard 1987, 20)

Facing the problem of truth, already critically addressed by Lessing in terms of historical truth that could found a salvation (an eternal beatitude, as Christianity claims), the author, by exploring the difference between Christ and Socrates, overturns the Socratic thesis that the truth is in man; according to the Danish thinker, man is rather in error, in sin, and only Jesus as Savior, Redeemer and Reconciler is the master.

It deals with a unique master, whom the disciple can no longer forget (note here the role of memory and oblivion) without falling back into bondage (Kierkegaard 1987, 69): being a disciple of Christ therefore entails not only receiving the truth from him (as in the Socratic relationship), but also the condition for receiving it (i.e. the disciple is in the sin and God manifests himself). This happens in the

https://doi.org/10.1515/9783111313610-006

contemporaneity with Christ, which provokes in the disciple a radical inner change, a conversion, a rebirth: the "historical fact" Jesus Christ, God in time, God made man, has in itself a metahistorical, absolute, transcendent elevation whereby it is the *necessary* truth that can ground the eternal salvation, and with the personal decision, in the "moment" (the meeting of time and eternity in the *instant*), through the leap of faith (where will and obedience are in play), man finds salvation in the unique event that is Christ.

> If the situation is to be different, then the moment in time must have such decisive significance that for no moment will I be able to forget it, neither in time nor in eternity, because the eternal, previously nonexistent, came into existence [blev til] in that moment. With this presupposition, let us now examine the relations involved in the question: Can the truth be learned? (Kierkegaard 1987, 27–28)

And so the believer is not simply a witness in the directness "but is a contemporary in the autopsy of faith" (Kierkegaard 1987, 157).

> And, now, the moment. A moment such as this is unique. To be sure, it is short and temporal, as the moment is; it is passing, as the moment is, past, as the moment is in the next moment, and yet it is decisive, and yet it is filled with the eternal. A moment such as this must have a special name. Let us call it: *the fullness of time*[1] (Kierkegaard 1987, 30–31).

In it and for it, something happens that reveals more about the master's situation and that of the disciple.

As a teacher, in his own way,

> But the god needs no pupil in order to understand himself, and no occasion can act upon him in such a way that there is just as much in the occasion as in the resolution. What, then, moves him to make his appearance? He must move himself and continue to be what Aristotle says of him, πάντα κινεί [unmoved, he moves all]. But if he moves himself, then there of course is no need that moves him, as if he himself could not endure silence but was compelled to burst into speech. But if he moves himself and is not moved by need, what moves him then but love, for love does not have the satisfaction of need outside itself but within. His resolution, which does not have an equal reciprocal relation to the occasion, must be from eternity, even though, fulfilled in time, it expressly becomes the moment, for where the occasion and what is occasioned correspond equally, as equally as the reply to the shout in the desert, the moment does not appear but is swallowed by recollection into its eternity. The moment emerges precisely in the relation of the eternal resolution to the unequal occasion. If this is not the case, then we return to the Socratic and do not have the god or the eternal resolution or the moment. Out of love, therefore, the god must be eternally resolved in this way, but just as his love is the basis, so also must love be the goal, for it would indeed

1 The biblical reference (Gal 4:4) emphasizing the Christological density of the moment category.

> be a contradiction for the god to have a basis of movement and a goal that do not correspond to this. The love, then, must be for the learner, and the goal must be to win him, for only in love is the different made equal, and only in equality or in unity is there understanding. Without perfect understanding, the teacher is not the god, unless the basic reason is to be sought in the learner, who rejected what was made possible for him[2] (Kierkegaard 1987, 37).

For his part, the disciple, receiving in the moment the condition (and not just the occasion) to be in truth and salvation, experiences a change in the direction of his journey, which is to be understood as conversion, repentance and rebirth (in the sense that he comes into the world again).

In all this, the paradox immediately arises. "But one must not think ill of paradox, because paradox is the passion of thought, and thought without paradox is like the lover without passion: a trivial protector"[3] (Kierkegaard 1987, 92). This paradox becomes apparent when it is necessary to confront the scandal of Christ, an absolute paradox that presents itself in the appearances of what immediately denies God as weakness, suffering and death.[4]

> Yet it has to be this way, and it is love that gives rise to all this suffering, precisely because the god is not zealous for himself but in love wants to be the equal of the most lowly of the lowly. When an oak nut is planted in a clay pot, the pot breaks; when new wine is poured into old leather bottles, they burst. What happens, then, when the god plants himself in the frailty of a human being if he does not become a new person and a new vessel! But this becoming – how difficult it really is, and how like a difficult birth! And the situation of the understanding – in its frailty, how close it is at every moment to the border of misunderstanding when the anxieties of guilt disturb the peace of love. And the situation of understanding – how terrifying, for it is indeed less terrifying to fall upon one's face while the mountains tremble at the god's voice than to sit with him as his equal, and yet the god's concern is precisely to sit this way (Kierkegaard 1987, 44–45).

---

2 "In relation to Him the historical detail does not have at all the importance it has for Socrates, because Christ is the Christ, an eternal present, being true God" (Kierkegaard 1962, 645).

3 The supreme paradox of thought lies in wanting to discover something that thought cannot think. It is a passion present in every expression of thought.

4 These aspects of Kierkegaard's dramatic Christianity are grounded in the necessity of the freedom of faith that is not determined by a direct vision of God. The paradox discussed here has nothing to do with Hegelian terminology, the dialectical movement or the theory of contradiction; rather, it expresses the Christian theology of the incarnation and the cross in such a way as to exalt the foundational option, the leap of faith that makes subjectivity happen hand in hand with the inability of demonstration to arrive at a final and existentially demanding conclusion. If anything, it approaches the positions of Luther of the *Dispute of Heidelberg* and *De servo arbitrio* concerning the manifestation of God *sub contraria specie, in opposito signo*.

This paradox testifies that beyond the inevitable but salutary objective uncertainty that reason can achieve, it is only in the infinite subjective passion, that is, in the interiority purified by choice not without, but beyond works, that the need for redemption opens up to the possibility of eternal salvation.

However, every scandal, by its very nature, is a misunderstanding of the moment, because the scandal is in front of the paradox and this, in turn, is the moment. Here, the eternal condition is given in time, but as long as the eternal and the historical face each other, the historical element is only an occasion; the paradox, on the other hand, reunites the contradiction, raises the historical reality (characterized by the freedom that presides over becoming) to the level of eternity and brings the eternal to the historical level, because the object of faith – faith that is not knowledge, but an act of freedom, an expression of the will and the sense of becoming – is not doctrine but the master.

2 The Eternity in the *Co-Belongingness* of Time and Being, Presence and Event (Heidegger)

In Martin Heidegger's *Denkweg*, there are abundant pages dedicated to time and space in their connection with Being-there (*Dasein*), the temporality, the question of Being (*Sein*) in its difference from entity (*Seiendes*) and the Event (*Ereignis*). Explicit textual references to the nexus of time and eternity are rare.

An interesting moment to test what becomes of our theme – time, eternity and freedom – is the so-called *kehre* of thought that stretches between *Being and Time* (unfinished masterpiece of 1927) and *Time and Being* (lecture of 1962): first, the magician of the Black Forest moves in search of Being within the fundamental experience of the oblivion of being; subsequently, moving from the Being of the Being there to the *Es* that *gibt* and from this to the *Ereignis*, he poses the question of what is proper to being and what is proper to time.

The second section of *Being and Time*, through the tracing back of the spatiality of Being-there (*Dasein*, which is already always in view of itself) to the ecstatic temporality (temporization) that characterizes the Care as Being-in-the-world, highlights the temporality as the ontological sense of the Care. From this point of view, the temporality does not mean what is ordinarily understood by time when speaking of space and time: it is not an entity, but it is temporalized and grasped at the conjunction between the original (Being-for-death, finiteness) and the authentic (established through the analysis of the stimulus of the moral consciousness that calls the Being to its most proper and authentic possibilities by speaking constantly in the mode of silence). Hence the temporality of beingness temporalizes a behavior that relates to time in such a way as to take it into account (counting, measuring, datability – now, then); from this arises the complex ordinary concept of time that also concerns history and natural events and does not coincide with that of temporality, which is its original concept. Therefore, worldly time is more objective than any possible object and more subjective than any subject insofar as it springs from being-in-the-world; it is not simply present in either the subject or the object because it is neither a condition of possibility. And it is with reference to this ordinary concept of time and the interplay of constant presence and change that characterizes it that we find a nod to eternity:

> Die Jetzfolge wird als ein irgendwie Vorhandenes aufgefasst; denn sie rückt selbst "in die Zeit". Wir sagen: *in* jedem Jetzt ist Jetzt, *in* jedem Jetzt verschwindet es auch schon. In *jedem* Jetzt ist das Jetzt Jetzt, mithin ständig *als Selbiges* anwesend, mag auch in jedem Jetzt je ein anderes ankommend verschwinden. Als *dieses* Wechselnde zeigt es doch zugleich die ständige Anwesenheit seiner selbst, daher denn schon *Platon* bei dieser Blickrichtung auf die Zeit als entstehend-vergehende Jetztfolge die Zeit das Abbild der Ewigkeit nennen musste[5] (Heidegger 2001, 423).

However, here – in the infinity of public time – the *escape* of being there from its authentical existence is announced, and this also concerns this way of thinking about the eternity.

In *Time and Being* (where it is declared that *Being and Time*'s attempt to bring the spatiality of being there back to temporality is no longer viable) the hitherto unthought-of sense of time that rests in *being as presence* (*Anwesen*) is recovered in the most original relation to the *Ereignis*.

> Was gibt den Anlaß, Zeit und Sein zusammen zu nennen? Sein besagt seit der Frühe des abendländisch-europäischen Denkens bis heute dasselbe wie Anwesen. Aus Anwesen, Anwesenheit spricht Gegenwart. Diese bildet nach der geläufigen Vorstellung mit Vergangenheit und Zukunft die Charakteristik der Zeit. Sein wird als Anwesenheit durch die Zeit bestimmt. Daß es sich so verhält, könnte schon genügen, um eine unablässige Unruhe in das Denken zu bringen. Diese Unruhe steigert sich, sobald wir uns aufmachen, dem nachzudenken, inwiefern es diese Bestimmung des Seins durch die Zeit gibt. Inwiefern? Dies fragt: Weshalb, auf welche Weise und woher spricht im Sein dergleichen wie Zeit? Jeder Versuch, das Verhältnis von Sein und Zeit mit Hilfe der landläufigen und ungefähren Vorstellungen von Zeit und Sein hinreichend zu denken, verstrickt sich alsbald in ein unentwirrbares Geflecht kaum durchdachter Beziehungen. Wir nennen die Zeit, wenn wir sagen: Jedes Ding hat seine Zeit. Aber ist das Sein ein Ding? Ist das Sein so wie ein jeweilig Seiendes in der

5 In the *Timaeus*, the nature of the soul is eternal, although it is not possible to fully bestow eternity on the generated so that he thinks of creating an image of eternity (αἰῶνος) and by ordering the heavens he creates of the eternity that remains in humanity an eternal image, which proceeds according to number, that which we have called time (cf. Plato, *Timaeus*, 37d).

> Zeit? Ist das Sein überhaupt? Würde es sein, dann müßten wir es unweigerlich als etwas Seiendes anerkennen und demzufolge unter dem übrigen Seienden als ein solches vorfinden. Dieser Hörsaal ist. Der Hörsaal ist beleuchtet. Den beleuchteten Hörsaal werden wir ohne weiteres und ohne Bedenken als etwas Seiendes anerkennen. Aber wo im ganzen Hörsaal finden wir das "ist"? Nirgends unter den Dingen finden wir das Sein. Jedes Ding hat seine Zeit. Sein aber ist kein Ding, ist nicht in der Zeit. Gleichwohl bleibt Sein als Anwesen, als Gegenwart durch Zeit, durch Zeithaftes bestimmt (Heidegger 1989, 371).

The *Es gibt* is first clarified in relation to giving (*das Geben*), then in relation to the *Es* that gives. The latter is interpreted as the appropriating-impropriating event (*das Ereignis*) in which the *lethe* of origins is resumed and from which emerges the abyssal foundation (the Being) in which the entities stand out in their difference from that. Hence the destiny and the gathering in the destinations that order the presence of being in its epochal moments. The giving in *Es gibt Zeit* is shown as the freeing of being in the Open (*das Offene*): in this, the word time no longer thinks one after the other of the succession of nows, but names the free-space of time (*Zeit-Raum*) where – with a reference to the Platonic *Chora* – the Open is thinned out and becomes free in the coming and going of the future, of having been and of the present. Authentic time thus conceals the giving of a giving in which the *Ereignis ereignet*.

The *Contributions to Philosophy* in a certain way deepen all this and show how space-time as understood in this way belongs to the truth of the essential occurrence of being as event. Here, moreover, in the oscillation of the event, a singular nod to eternity takes shape as that which can return not as the same but as that which is transformed anew, being such that it is recognized in this manifestation not as the same.

> Das Ewige ist nicht das Fort-währende, sondern jenes, was im Augenblick sich entziehen kann, um einstmals wiederzukehren. Was wiederkehren kann, nicht als das *Gleiche*, sondern als das aufs neue Verwandelnde, Eine-Einzige, das Seyn, so dass es in dieser Offenbarkeit zunächst nicht als das Selbe erkannt wird! Was ist dann *Ver-ewigung*? (Heidegger 2007, 6–7)

This question of making eternal, which we leave open here, opens up a new perspective in relation to *Being and Time* in order to understand eternity in relation to time-space, connected to perpetuity, to the moment, to transformation.

From what has been said, two perspectives on the eternal emerge in Heidegger, albeit only in the manner of hints, one in *Being and Time*, the other in the *Contributions*. Both are to be set in the Heideggerian *Denkweg* where, unlike in Hegel, finiteness is made visible, that of man and of the *Ereignis* itself. This finiteness (different from when it is understood in relation to infinity) is finiteness in itself and at the same time is in some way related in both cases – albeit only in the indicative manner of the hint – to the eternal/eternity so that this relationship concerns and dimensions it: with regard to beingness, it says that it, marked by being-for-death, exists finitely and at the same time is to be understood as the original finiteness of temporality, which does not deny that time "continues to flow" and only indicates the phenomenal character of the original temporality that is revealed in what is projected in the existential project of Being itself; in relation to the Event and man in relation to it, on the other hand, it indicates the limit, its own, lodged in one's own, which, starting from the *Ereignis* itself, from the idea of appropriateness as the intimate appropriate force proper to it, reveals itself and it hides. In the first case, the temporal eternity is the image of another eternity, elsewhere and unique, which is imagined in the concealment of inauthentic existence; in the second case, it is what is subtracted in the moment in order to return.

3 Time *in Direction* of Eternity (Ricœur)

In Paul Ricœur's trilogy *Time and Narrative*, the nexus of time and eternity lies within the reflection on the possibilities and limits of the attempt to express time realized by highlighting how the possibility at stake in the structural identity of the narrative function and its truth lies in the temporal nature of human existence. Indeed, in the invention of the intrigues that weave the narrative lies a work of synthesis where ends, causes and causality are brought together in the temporal unity of a total and complete action that is of great help in the arduous task of explaining what time is. The result is that temporality cannot be expressed in the direct discourse of phenomenology or cosmology and requires the direct discourse of narration: it is a question of considering the tale as the custodian of time insofar as there would be no time thought if it were not narrated. It is a question of considering the tale as the custodian of time insofar as there would be no time thought if it were not narrated. We must therefore overcome the dead end that opts for the attempt to derive the principle of the measurement of time from the mere relaxation of the spirit (Augustine) or from the mere movement of nature, from the world, from the universe (Aristotle), in order to recognize that a psychological and a cosmological theory of time conceal each other to the extent that they implicate each other reciprocally. Here, the fact that psychology is legitimately added to cosmology, without being able to dislocate it and without either of them, taken separately, proposing a satisfactory solution to their unbearable disagreement, constitutes the aporetic side of the question.

With regard to the distension of the spirit, Ricœur refers to Augustine, who in the *Confessions* starts from the antithesis between *intentio* and *distensio animi* interpreted from Gen 1:1. Augustine's analysis of time in connection with these analyses has a strongly interrogative, indeed aporetic, character, which none of the ancient theories of time, from Plato to Plotinus, have brought to such a level of radicality. The question in fact sounds: *quid est enim tempus*? A question that underlies all the ancient difficulties about the being and non-being of time. In posing the question, the North African thinker also poses an open question, because while skeptical argumentation inclines towards non-being, the confidence measured in the everyday use of language obliges us to say, albeit in a way that is not yet clear, that time is.[6] But, if it is true that we speak of time in a sensible way and in positive terms, it is equally true that in Augustine, it is precisely from this certainty that the inability to explain how the notion of *distensio animi*, combined with *intentio*, relates to the question of the measurement of time already raised by Aristotle and the ancient thinkers arises. In this regard, the language is in fact limited to the *fact* of measurement, while the *how* eludes him. This is how the Hipponate seeks by studying the life of the human soul.

> Augustin va d'abord paraître tourner le dos à la certitude que c'est lepassé et le futur qu'on mesure. Ultérueurement, en mettant le passé et le futur dans le présent, par le biais de la mémoire et de l'attente il pourrasauver cette certitude initiale d'un désastre apparent, en transférant sur l'attente et sur la mémoire l'idée d'un long futur et d'un long passé. Mais cette certitude du langage, de l'expérience et de l'action ne sera recouvrée qu'aprés avoir été perdue et profondément transformée (Ricœur 1983, 23).

The question of *how* develops at this point like the question of *where*: that is, it is a question of seeking a place for future and past things as they are recounted and foretold, and this place turns out to be the human soul. Here, narration implies memory (the imprint left by events that remains fixed in the spirit) and the expected prediction (a presentiment that allows us to foretell, sign and cause future things that are thus present and anticipated). An elegant solution, but precarious: in it, the structure of an image that is now a footprint of the past, now a sign of the future, is a problem. Only by confronting the enigma of the mystery of time can Augustine overcome this precariousness.

In order to solve the enigma, Augustine sets aside the cosmological solution so as to force the investigation to search exclusively in the soul and thus in the multiple structure of the threefold present that underlies extension and measurement. In all this for Ricœur

6 The skeptical argument is well known: time has no being because the future is not yet, the past is no more and the present does not remain. Yet we speak of time as belonging to being: we say that things to come will be, that things past have been, and that things present pass (and passing is nothing).

> La trouvaille inestimable de saint Augustin, en réduisant l'extension du temps à la distension de l'âme, est d'avoir lié cette distension à la faille qui ne cesse de s'insinuer au coeur du triple présent: entre le présent du futur, le présent du passé et le présent du présent. Ainsi voit-il la *discordance* naître et renaître de la *concordance* même des viséès de l'attente, de l'attention et de la mémoire. C'est à cette énigme del la spéculation sur le temps que répond l'acte poétique de mise en intrigue (Ricœur 1983, 41).

It is at this point that eternity comes into play in the *Confessions* because something is still missing from the full sense of *distensio animi* and this can only emerge from the contrast with eternity. What is missing is not the sufficient sense of *distensio* that is sufficient to respond to the aporias of non-being and measure. What is missing is of a quite different order.

Hence, Ricœur indicates the three main incidences of Augustine's meditation on eternity over time: the first locates the discourse on time within the horizon of an idea-boundary that forces us to think about time and the other from time together; the second operates on the existential level and intensifies the experience of *distensio*; the third calls this same experience to overcome itself in the direction of eternity and thus to inwardly hierarchize itself against the facet for the representation of a linear time. All of this emerges in relation to the biblical context of the eternity of the Word (which insinuates a *creatio ex nihilo* and is not satisfied with demiurgy) on the basis of which the Hipponate attributes to the eternal Word an eternal reason that assigns things a beginning and an end, leading to an idea of eternity as *semper stans* inwardly moved by its Trinitarian life (stability in becoming, instability of things).

> La nègativitè est ici à son comble: pour penser jusqu'au bout la *distensio animi*, c'est-à-dire la faille du triple présent, il faut pouvoir la "comparer" à un présent *sans* passé *ni* futur. C'est cette extréme négation qui sous-tend la réponse à l'argument d'apparence frivole (Ricœur 1983, 46).

This reality is best understood by taking into account that

> La conjonction de l'hébraïsme et du platonisme dans l'interprétation de l' *ego sum qui sum* d'Exode 3,14 dans sa traduction latine nous interdit d'interpréter la pensée de l'éternité comme une pensée sans object. En outre la conjonction de la louange et de la spéculation atteste qu'Augustin ne se borne pas à penser l'éternité; il s'addresse à l'Éternel, il l'invoque à la seconde personne: *sum* et non *esse* (Ricœur 1983, 47–48; Hadot 1987, 34–46).

From this perspective, thought has the onerous task of forming the idea of timelessness in order to think time as passage, as transiting, that is, time in the direction of eternity.

With this, Augustine, who provided a significant solution to the problem of the relationship between soul and time left unresolved by Aristotelianism, did not reject Aristotle's essential theory that gives movement priority over time. By stating that time is the measure of movement according to the before and after, the Stagirite makes it possible to observe: that time is relative to movement without being confused with it; the application to time of the relationship between before and after; its completion through the numerical relationship (the numerable number before the numerate number). Although there is no explicit reference to the soul in this (despite the references to the operations of perception, discrimination and comparison that cannot be other than those of the soul) Aristotle himself recognizes that there is the awkward problem of knowing whether or not time exists without the existence of the soul. In the Stagirite, however, it remains rather overshadowed by the rooting of change and motion in the *physis* that is its principle and cause.

The answer to the aporias that the reconciliation of the two visions – Augustinian and Aristotelian, psychological and cosmological, with all their expansions in the history of Western thought – includes is found by Ricœur in the poetics of time, which he constructs by confronting recounting (*mithos* [intrigue]), the triple *mimesis* (imitation and representation of action) and the narrative identity of the individual and the community with its own frailties. Narrative time – and with it the reference to historicity and history – is here like a bridge between the other two perspectives on time. It is in fact the elaboration of a third time.

> Ma thèse est ici que la manière unique dont l'histoire répond aux apories de la phénoménologie du temps consiste dans l'élaboration d'un *tiers-temps* – le temps proprement historique –, qui fait médiationentre le temps vécu et le temps cosmique. Pour démontrer la thèse, on fera appel aux *procédures de connexion*, empruntées à la pratique historienne ellemême, qui assurent *la réinscription du temps vécu sur le temps cosmique*: calendrier, suite des générations, archives, document, trace (Ricœur 1985, 147).

Ricœur is also aware that the proposed solution does not provide a fully adequate response to the aporia of temporality. But it is not an inadequacy that constitutes a check, both insofar as it makes the aporias of temporality and its links with the eternal work by making them productive, and because the recognition of the limits of its scope is what every theory is accomplished in.

4 Landings and Openings

The oscillation between the existential and cosmological conception of time that we have seen emerge from the confrontation of philosophical, theological, scientific and narrative elements demonstrates the aporetic character that introduces thought to certain figures of eternity: the moment as the relationship of an eternal decision with a non-proportional occasion; the finitude of being-for-death and/or that which in the moment can subtract itself in order to be able to return; the absence of time (*semper stans*) that allows one to think of time as passage. In these figures shines the freedom that in Kierkegaard shows itself as the decision of the leap of faith, in Heidegger as an anticipatory decision and as leave-to-be, in Ricœur as inventive capacity operating in the making of the intrigue of action. The renewed figures of time and eternity – as attested by the linguistic reserve in which they are spoken – if, on the one hand, they manifest the complexity, ambivalence and insufficiency of these categories and explanations, on the other hand, they nourish the vital restlessness that constantly characterizes the turning of thought towards its question.

References

Hadot, Pierre. 1987. *Platonismo e Idealismo*. Bologna: Il Mulino.
Heidegger, Martin. 1989. "Beiträge zur Philosophie (Vom Ereignis)". In *Gesamtausgabe*, Vol. 65. Frankfurt am Main: Vittorio Klostermann.
Heidegger, Martin. 2001. *Sein und Zeit*. Tübingen: Max Niemeyer.
Heidegger, Martin. 2007. "*Zeit und Sein*". In *Gesamtausgabe*, Vol. 14: *Zur Sache des Denkens*, 3–30. Frankfurt am Main: Vittorio Klostermann.
Kierkegaard, Søren. 1962. *Diario*. Edited and translated by Cornelio Fabro. Brescia: Morcelliana.
Kierkegaard, Søren. 1987. *Philosophical Fragments. Johannes Climacus*. Edited and translated by Howard V. Hong and Edna H. Hong. Princeton/NJ: Princeton University Press.
Ricœur, Paul. 1983. *Temps et récit*, Vol. 1: *L'Intrigue et le récit historique*. Paris: Éditions du Seuil.
Ricœur, Paul. 1985. *Temps et récit*, Vol. 3: *Le temps raconté*. Paris: Éditions du Seuil.

Second Part: **The Eternity Concealed in the Cosmos and the Secrets of Consciousness**

Roger Penrose

The Basic Ideas of Conformal Cyclic Cosmology

Abstract: The Conformal Cyclic Cosmology (CCC) proposes that the universe undergoes repeated cycles of (accelerated) expansion, named "aeons", where the maximal (or infinite) extension of the previous cycle goes to coincide with the Big Bang stage of the successive cycle. No contraction (big crunch) is required in this model. This is made possible through the conformal structure that dominates space-time at the beginning and at the end of each aeon. The CCC solves the paradox of the super-special initial conditions required by Second Law at the Big Bang, and among its observational consequences, predicts the presence of "circular rings" in the temperature fluctuations of the Cosmic Microwave Background spectrum.

1 Conformal Space-Time Geometry

In this article, I outline a new cosmological proposal – *Conformal Cyclic Cosmology*, or CCC – according to which the universe undergoes repeated cycles[1] of expansion, that I refer to as *aeons*, each starting from its own "big bang" and finally coming to a stage of accelerated expansion which continues indefinitely (which would be for an infinite time, according to how a clock made of physical material would measure time), in close accordance with current observations[2] of our own aeon. There is no stage of contraction (to a "big crunch") in this model. Instead, each aeon of the universe, in a sense "forgets" how big it is, both at its big bang and in its very remote future where it becomes physically identical with the big bang of the next aeon, despite there being an infinite scale change involved, on passing from one aeon to the next.

To put this in more mathematical terms, we must consider the geometrical structure that becomes of relevance at the cross-over from aeon to aeon. This is

1 The basic idea of CCC first came to me in August 2005. Comparison should be made with the proposals of Veneziano (1998) and of Steinhardt and Turok (2002, 2007). However, these others schemes differ substantially from CCC and depended upon ideas taken from string theory.
2 See Perlmutter et al. (1998).

Article note: Sections 2, 3, 4, and 5 of this article are taken almost *verbatim* from Penrose (2006), with the kind permission of the publishers.

https://doi.org/10.1515/9783111313610-007

what is referred to as *conformal* structure rather than a metric structure. Conformal structure may be viewed as family of metrics that are equivalent to one another via a *scale change*, which may vary from place to place. Thus, in conformal space-time geometry, we do not have any particular metric g_{ab}, but an equivalence class of metrics where the metrics $\hat{g}_{ab}$ and g_{ab} are considered to be equivalent if there is a smooth positive scalar field Ω for which $\hat{g}_{ab}=\Omega^2 g_{ab}$. A more directly physical way of putting this is to say that, for a space-time, its conformal structure is simply its *light-cone* structure. It will be appropriate to incorporate the distinction between *future* and *past* light cones into the notion of "conformal structure", this structure being then equivalent to the space-time's *causal structure.*[3] We may note that most of the information in the metric is actually in the conformal (or causal) structure, because only 1 out of the 10 components (per point) of the metric is needed to fix the scale, whereas the remaining 9 (as ratios) are what fix the locations of the light cones. It is this conformal structure which is to remain smooth at the cross-over from one aeon to the next.

The reader may wonder why one should consider it necessary to adopt this strange-seeming view of the history of the universe. There are, in fact, several different ingredients which point the way to such a proposal, various aspects of which are implicit in the nature of the cosmic microwave background (CMB). It has become customary – essentially part of what is now referred to as the "standard" picture of cosmology – to try to explain these in terms of the concept of "inflation", whereby a substantial period of exponential expansion it taken to be part of the universe's extremely early history.[4] However, inflation does not remotely come to terms with what I would regard as the most fundamental conundrum of all, concerning the extraordinarily special nature of the Big Bang,[5] and it is this that provided the initial motivation underlying CCC.

2 The Basic Conundrum

Proposals for describing the initial state of the universe hardly ever address a certain fundamental issue – yet this is an issue whose significance is, in a certain sense, obvious. This arises from one of the most fundamental principles of physics: the *Second Law of Thermodynamics*. According to the Second Law, roughly

3 See, for example, Kronheimer and Penrose (1967).

4 See Guth (1981).

5 I am using the capitalised form "Big Bang" for the specific event that initiated our aeon; "big bang" is used here more generally for the corresponding event for any aeon.

speaking, the entropy of the universe increases with time, where the term "entropy" refers to an appropriate measure of *disorder* or lack of "specialness" of the state of the universe. Since the entropy increases in the future direction of time, it must decrease in the past time-direction. Accordingly, the initial state of the universe must be the most special of all, so any proposal for the actual nature of this initial state must account for its extreme specialness. Proposals have been put forward from time to time (such as in various forms of inflationary cosmology and the previously popular "chaotic cosmology"[6]) in which it is suggested that the initial state of the universe ought to have been in some sense "random", and various physical processes are invoked in order to provide mechanisms whereby the universe might be driven into the special state in which it appears actually to have been in, at slightly later stages. But "random" means "non-special" in the extreme; hence the conundrum just referred to.

Sometimes theorists have tried to find an explanation via the fact that the early universe was very "small", this smallness perhaps allowing only a tiny number of alternative initial states, or perhaps they try to take refuge in the *Anthropic Principle*, which would be a selection principle in favour of certain special initial states that allow the eventual evolution of intelligent life. Neither of these suggested explanations gets close to resolving the issue, however.[7] It may be seen that, with time-symmetrical dynamical laws, the mere smallness of the early universe does not provide a restriction on its degrees of freedom. For we may contemplate a universe model in the final stages of collapse. It must do *something*, in accordance with its dynamical laws, and we expect it to collapse to some sort of complicated space-time singularity, a singularity encompassing as many degrees of freedom as were already present in its earlier non-singular collapsing phase. Time-reversing this situation, we see that an initial singular state could also contain as many degrees of freedom as such a collapsing one. But in our actual universe, almost all of those degrees of freedom were somehow not activated.

What about the Anthropic Principle? Again, this is virtually no help to us whatever in resolving our conundrum. It is normally assumed that life had to arise via complicated evolutionary processes, and these processes required particular conditions, and particular physical laws, including the Second Law. The Second Law was certainly a crucial part of evolution, in the way that our particular form of life actually came about. But the very action of this Second Law tells us that however special the universe may be now, with life existing in it now, it must have been far *more* special at an earlier stage in which life was not present.

---

6 See Misner (1968), for example.

7 Penrose (1990, 2004, Chapter 27).

From the purely anthropic point of view, this earlier far more special phase was not needed; it would have been much more likely that our present "improbable" stage came about simply by chance, rather than coming about via an earlier even more improbable stage. When the Second Law is a crucial component, there is always a far more probable set of initial conditions that would lead to this same state of affairs, namely one in which the Second Law was *violated* prior to the situation now!

As another aspect of this same issue, we may think of the vastness of our actual universe, most of which had no actual bearing on our existence. Though very special initial conditions were indeed required for our existence in our particular spatial location, we did not actually need these same special conditions at distant places in the universe. Yet as we look out at the universe, we see the same kind of conditions, acting according to the same Second Law of Thermodynamics, no matter how far out we look. If we take the view that the Second Law was introduced in our vicinity merely for our own benefit, then we are left with no explanation for the extravagance of this same Second Law having to be invoked uniformly throughout the universe, as it appears to be as far as our powerful instruments are able to probe.

3 The Enormity of the Specialness

In order to stress the extraordinary scale of this problem, and the intrinsic implausibility of explanations of this kind, it is helpful to enter a little more precisely into the definition of entropy, and to estimate the entropy magnitudes that we have to contend with. Boltzmann provided us with a beautiful formula for the entropy S of a system:

$$S = k \log V$$

Here k is Boltzmann's constant and V is the volume of a certain region in the total *phase space*[8] $\mathcal{P}$ of the system under consideration. We are taking $\mathcal{P}$ to be "coarse-grained" into sub-regions, each sub-region representing states that are deemed to be indistinguishable with regard to any reasonable macroscopic parameter. (There is clearly an element of arbitrariness or subjectivity, here, as to which parameters

8 Phase space is a mathematical space $\mathcal{P}$, normally of an enormous number of dimensions, in terms of which a single point of $\mathcal{P}$ encodes the entire configuration of a physical system together with the instantaneous motions of all its parts (given in terms of their individual positions and momenta); see, for example, Penrose 2004, Chapter 20.

are to be regarded as macroscopically discernible and which are deemed to be effectively "unmeasurable". In practice, there is a considerable robustness with regard to this arbitrariness, and it is reasonable to disregard this issue in the present discussion.) Any particular state of the system under consideration will be specified by some point x of $\mathcal{P}$, and the quantity V is then the volume of the particular sub-region of $\mathcal{P}$ which contains x.

With regard to future time-evolution of the system, the Second Law can be understood as the fact that, as the system evolves, the point x moves within $\mathcal{P}$ so that with overwhelming probability it enters sub-regions of successively larger and larger volume V. This arises from the fact that, in practice, the sub-regions differ stupendously in size. The logarithm in Boltzmann's formula helps here (as does the smallness of k, in ordinary units), because there need only be a modest increase in S when x moves from one sub-region into a neighbouring one of stupendously larger volume. But this is only the easy half of our understanding of the Second Law. The difficult half is to understand why, when we *reverse time*, x enters successively *tinier* sub-regions of $\mathcal{P}$. It does this because it has ultimately to reach the exceptionally tiny region $\mathscr{B}$ which represents the Big Bang itself. The difficult half of the Second Law involves an understanding of why the universe had to start off in such an extraordinarily special state. And to understand how special the Big Bang actually was, we need to compare the volume of $\mathscr{B}$ with that of the entire phase space $\mathcal{P}$.

One point of concern is the fact that the entire volume might be infinite, as it certainly would be in the case of a spatially infinite universe. This issue, while of relevance, is not of major importance for our considerations here. There is also the issue of how we get a finite phase-space volume when some of the parameters would be describing continuous fields. I shall evade this latter issue by assuming that it is dealt with by quantum mechanics, where for a finite universe of bounded energy content we may assume only finitely many quantum states.

To deal with a spatially infinite universe, I shall assume that we need consider only, say, that *co-moving*[9] portion of the universe that intersects our past light cone. This contains something of the order of 10^{80} baryons. To obtain a lower bound for the volume of $\mathcal{P}$, for this situation, we can consider the entropy that arises when this number of baryons is collapsed into a black hole. For this, we use the Bekenstein-Hawking entropy formula[10] $S_{BH}=8\pi^2 kGm^2/hc$ for a spherical black hole of mass m and find a value of the order of 10^{123}. If we envisage the

9 In cosmology, a "co-moving" region refers to a part of space-time that follows the history of the motions of the (idealised) galaxies; see, for example, Rindler (2001).

10 See Hawking (1976a). Here k is Boltzmann's constant, G is Newton's gravitational constant, h is Planck's constant, and c is the speed of light.

dark matter also being absorbed into this black hole, we would get a considerably larger entropy (and, for a continually expanding universe, we must contemplate possibly even larger entropy values that might be attained in the very remote future), but this value represents a usable lower bound. Recalling the logarithm in Boltzmann's formula (this being a natural logarithm, but that is of no concern), we find that the volume of $\mathcal{P}$ is greater than that of $\mathcal{B}$ by a factor that exceeds

$$10^{10^{123}}$$

This gives us some idea of the enormity of the precision in the Big Bang!

4 The Geometric Nature of the Specialness

A seeming paradox arises from the fact that our best evidence for the very existence of the Big Bang arises from observations of the microwave background radiation – frequently referred to as the "flash of the Big Bang", greatly cooled down to its present value of ~2.7K. The intensity of this radiation, as a function of frequency, matches the Planck radiation formula extraordinarily closely, giving us impressive evidence of an early universe state with matter in *thermal equilibrium*. But thermal equilibrium is represented, in phase space ***P***, as the coarse-graining sub-region of largest volume (so large that it normally exceeds all others put together). This corresponds to *maximum* entropy, so we reasonably ask: how can this be consistent with the Second Law, according to which the universe started with a very *tiny* entropy?

The answer[11] lies in the fact that the high entropy of the microwave background refers only to the *matter* content of the universe and not to the *gravitation field*, as would be encoded in its space-time geometry in accordance with Einstein's general relativity. What we find, in the early universe, is an extraordinary *uniformity*, and this can be interpreted as the gravitational degrees of freedom that are potentially available to the universe being not excited at all. As time progresses, the entropy rises as the initially uniform distribution of matter begins to clump, as the gravitational degrees of freedom begin to be taken up. This allows stars to be formed, which become much hotter than their surroundings (a thermal imbalance that all life on Earth depends upon), and finally this gravitational clumping leads to the presence of black holes (particularly the huge ones in galactic centres), which represent an enormous increase in entropy.

11 See, for example, Penrose (2004, Chapter 27).

Although, in general, there is no clear geometric measure of the entropy in a gravitational field in general relativity, we can at least provide proposals for the *non-activation* of gravitational degrees of freedom at the Big Bang. I have referred to such a proposal as the *Weyl Curvature Hypothesis*[12] (WCH). In Einstein's theory the *Ricci curvature* R_{ab} is directly determined by the gravitational sources, via the energy-momentum tensor of matter (analogue of the charge-current vector J^a in Maxwell's electromagnetic theory) and the remaining part of the space-time Riemann curvature, namely the *Weyl curvature* C_{abcd}, describes gravitational degrees of freedom (analogue of the field tensor F_{ab} of Maxwell's theory). WCH – which is a time-*asymmetrical* hypothesis – asserts that *initial* space-time singularities must be constrained to have $C_{abcd}=0$ (in some appropriate sense), whereas final space-time singularities (as occur inside black holes) are unconstrained.

What appears to be the most satisfactory form of WCH has been studied extensively by Paul Tod.[13] This proposes that an initial space-time singularity can always be represented as a smooth *past boundary* to the *conformal geometry* of space-time. Tod's formulation of WCH is the hypothesis that we can adjoin a (past-spacelike) hypersurface boundary to space-time in which the conformal geometry can be mathematically extended smoothly through it, to the past side of this boundary. This amounts to "stretching" the metric by a conformal factor Ω which becomes *infinite* at the Big Bang singularity, so that we get a smooth metric $\hat{g}_{ab}$ which actually extends across this boundary.

5 Conformal Cyclic Cosmology

So far, we regard the conformal "space-time" prior to the Big Bang as a mathematical fiction, introduced solely in order to formulate WCH in a mathematically neat way. However, CCC takes this mathematical fiction seriously as something *physically real*. But what "physical reality" can we consistently attach to this space-time occurring "before the Big Bang"? As a clue to this possibility, we should consider the nature of the physics that is presumed to be taking place just *after* the Big Bang. (I am ignoring the possibility of inflation here, as CCC seems to provide an alternative explanation for the main positive achievements of inflation; see Section 6.) As we approach the Big Bang, moving back in time, we expect to find temperatures that are increasingly great. And the greater the temperature, the more irrelevant the rest-masses of the particles involved will become, so

12 Penrose (1979).
13 Tod (2003).

these particles are effectively massless near the Big Bang (presumably at energies that were greater than that provided by the mass of the Higgs particle). Now, massless particles (of whatever spin) satisfy conformally invariant equations.[14] I am going to suppose that the *interactions* between these massless entities are also described by conformally invariant equations. (This seems to be consistent with current understanding of particle physics.) With such conformal invariance holding in the very early universe, the universe has no way of "building a clock". So it loses track of the scaling which determines the full space-time metric, while retaining its conformal geometry.[15]

We may apply considerations of this kind also to the distant future of the universe. If we assume that in the very remote future, conformally invariant equations again govern the universe's contents, then we can apply the same mathematical trick as before, but now in the reverse sense that we look for a boundary at which the conformal factor Ω becomes *zero*, rather than infinite. This amounts to using a metric, such as $\hat{g}_{ab}$ above, in which the future infinity is "squashed down" to be a finite boundary to space-time, which is conformally regular in the sense that the space-time can be mathematically extended across this future boundary as a smooth conformal manifold.[16] If we also assume that there is a positive cosmological constant present, as current observations appear to point strongly towards,[17] then we find that this future conformal boundary is *spacelike*.

There is, however, a crucial difference between the use of a conformal boundary to study the future asymptotics of a space-time and Tod's use of a conformal boundary to treat the Big Bang. For in the latter case the very validity of this trick provides a formulation of WCH, whereas in the future situation of an expanding universe with conformally invariant contents, the validity of this procedure is more or less *automatic*.[18] Physically, we may think that again in the very remote future, the universe "forgets" time in the sense that there is no way to build a clock with just conformally invariant material. This is related to the fact that massless particles, in relativity theory, do not experience any passage of time. We might even say that to a massless particle, "eternity is no big deal". So the future boundary, to such an entity is just like anywhere else. With conformal invariance both in the remote future and at the Big Bang origin, we can try to argue that the two situations are *physically identical*, so the remote future of one aeon of the universe becomes the Big Bang of the next. This is the basis of CCC.

---

14 Penrose (1965).
15 See also Rugh and Zinkernagel (2007).
16 See Penrose (1965).
17 For example, Perlmutter et al. (1998).
18 This is a reasonable conclusion from the work of Friedrich (1986).

6 Physical Implications

There are certain important assumptions involved in CCC, in order that only conformally invariant entities survive to eternity. One of these is that black holes will all eventually evaporate away and disappear. This evaporation is a consequence of Stephen Hawking's quantum considerations,[19] and these are now normally accepted. There is the issue of whether black holes would actually ultimately disappear or perhaps leave some form of "remnant". I am here taking the more conventional view that they would indeed disappear in a final (cosmologically very mild) explosion. However, there is another aspect to this concerning the so-called "information paradox", and I shall need to return to this in the last section.

A more immediate issue, for CCC, is how to get rid of massive fermions and massive charged particles. It is not too unconventional to assume that protons will ultimately decay, or even that there could be one variety of neutrino that is massless, but the real problem lies with electrons. A good many of them will annihilate with positively charged particles, but there will be a relatively small number of "stray" charged particles (electrons, positrons, and also perhaps protons if these do not decay) which become trapped in their ultimate cosmological event horizons, being unable to come in contact with other particles of opposite charge. There are various possible ways out of this, none of which is part of conventional particle physics. One possibility is that electric charge is not exactly conserved, so that within the span of eternity, electric charge would eventually disappear. A much more satisfying possibility, is that *rest-mass* is not, in a sense, completely constant, so that the electron's mass will eventually decay away – and, indeed, so would the rest-mass of all massive particles (e.g. neutrinos), perhaps at different rates, albeit extremely slowly, over the infinitude of eternity. This seems to me to be not at all implausible, because the very reason that rest-mass is a "good quantum number" is the result of it being a *Casimir operator* for the Poincaré group (i.e. commuting with all the generators of the group). However, when there is a cosmological constant present (as is crucial for CCC in any case), the Poincaré group is not completely correct for understanding particle physics, and we need to turn to the deSitter group for something more appropriate. The rest-mass is *not* a Casimir operator for the deSitter group, so it may not be implausible for there to be a very slow decay of rest-mass over enormous stretches of time It should be made clear that I do not mean the decay of massive particles into massless ones, but a universal decay of rest-mass itself, which is not necessarily at the

---

19 It follows from quantum field theoretic considerations of pair annihilation that there cannot be any massless charged particles around today. See Bjorken and Drell (1965).

same rate for all types of particle. Accordingly, charged particles such as electrons and positrons would asymptotically[20] lose their mass, so that in the asymptotic limit of the remote future, all mass is lost, so conformal physics again becomes the relevant physics, and conformal geometry the relevant space-time geometry, just as it had been at the Big Bang.

These matters are all tied up with the issue of the strength of the gravitational interaction, which I have postponed in my discussion here. In the background of conformal geometry, the strength of gravity may be considered as being infinitely large at the Big Bang (which is, in a sense, why the gravitational degrees of freedom must initially be set to zero), and this strength gets smaller as time progresses, eventually reducing to zero at the final boundary. To express all this in a satisfactory mathematical framework for CCC, we need to reformulate general relativity in an appropriately conformally invariant way. This can indeed be done. We take advantage of the fact that the Weyl tensor $C_{abc}{}^{d}$ is conformally invariant, and provides a precise measure of the conformal curvature of space-time. We can define the gravitational "spin-2 field" $K_{abc}{}^{d}$ to be described by $C_{abc}{}^{d}$ with respect to the original space-time metric g_{ab}, but when we pass to the conformally related metric $\hat{g}_{ab}=\Omega^2 g_{ab}$ we find that, curiously, $K_{abc}{}^{d}$ picks up a factor of Ω^{-1}, which $C_{abc}{}^{d}$ does not.[20]

This has the implication that gravitational radiation (described by $K_{abc}{}^{d}$) actually survives at the future boundary (whereas $C_{abc}{}^{d}$ vanishes there) and its presence shows up as a non-zero normal derivative of $C_{abc}{}^{d}$ at the boundary. This gives rise to primordial density and velocity fluctuations at the Big Bang. The details of all this have yet to be fully worked out, but, in principle at least, there should be clear-cut predictions which should be observable. Some of these will be discussed in the next section.

7 Observational Effects in the CMB

There are numerous observational consequences of CCC, the most immediate of which would appear to be the way in which it should lead to temperature variations in the cosmic microwave background. The observed variations are at the level of only about 1 part in 10^5. The presently conventional view of how these come about is via quantum fluctuations at the Big Bang becoming stretched out by inflation to cosmological scales. The exponential – and essentially scale-invariant – expansion, that is taken to be present in the inflationary phase, would provide a

20 See Penrose (1965).

distribution of density variation, and hence temperature variation, that is scale-invariant, this comparing rather favourably with detailed CMB observations over a broad spectrum of sizes. This scale invariance is regarded as a success of inflationary theory, but in CCC there is no inflation, as such. But it would appear that CCC has implications that are rather similar, in this respect, since it also involves an exponential expansion, though this occurs *before* the Big Bang in CCC, rather than afterwards.

The origin of the irregularities, according to CCC would have to be something quite different from quantum fluctuations, however. It may be that irregularities in the ultimate energy distribution in the previous aeon would simply propagate into the next aeon. But there is a particular factor which could provide a very significant and interesting input into the energy and velocity distribution of the primordial material in the subsequent aeon. This would be the effect of encounters between supermassive black holes that inhabit the centres of galaxies. These would be expected to occur at the later stages of the aeon previous to ours, and the gravitational waves emitted in such encounters could well carry away an appreciable proportion of the mass content of the black holes themselves in a relatively brief burst of gravitational energy. Owing to effects discussed at the end of Section 5, these bursts ought to give rise to circular rings along which the CMB temperature should appear either slightly warmer or slightly cooler than the average CMB temperature, depending mainly upon whether the source is very distant or comparatively close, as compared with the "particle horizon" of our current aeon. The effect ought to be somewhat similar to the pattern of ripples on a pond following a period of rain which had recently stopped: each raindrop produces a circular ring, but a detailed statistical analysis would be required in order to ascertain whether the resulting pattern of ripples is composed in this way. A preliminary search by Amir Hajian, under the direction of David Spergel were not adequate to see this effect. However, later analyses by Vahe Gurzadjan and by a Polish group led by Krzysztof Meissner gave powerful evidence for the existence of this signal. The Polish group giving a confidence level of around 99.5%. There is also another predicted effect of CCC namely that which I refer to as Hawking Points. In the previous aeon most of the matter will have been in galactic clusters, each of which will ultimately become largely swallowed by a supermassive black hole. As that aeon evolves the temperature gets colder as it expands until it becomes colder even than the Hawking temperature of such black holes. At that point those black holes will begin to evaporate by Hawking radiation, but since this happens only in the very late stages of that aeon (something like 10^{100} years) the entire radiation reaches the conformal infinity of the aeon in a very tiny point-like region. The mass-energy of that cluster comes through into our aeon at what is essentially a single point and that is what I refer to as a Hawking Point. Subsequently, in our aeon, the energy in the Hawking Point spreads out over a region until, after 380,000 years the energy of the Hawking radiation spreads out to the region, visible in the CMB as a spot which turns out to be about 8 times the diameter of the full moon. Such spots are observed in the CMB at a confidence level of 99.98% according to an analysis described in a paper by Daniel An, Krzystztof Meissner, Pawel Nurowski, and myself.

8 The Second Law of Thermodynamics

We must return, finally, to the issue of the Second Law of Thermodynamics, which initiated this entire line of thinking. Although the very structure of CCC demands a very particular structure for the big bang of every aeon, there is a conundrum still remaining. For in matching the aeon's future conformal surface, which represents what we expect to take as a very high entropy state representing the remote future of any particular aeon, with the very low entropy big-bang state of the succeeding aeon, we find the seeming contradiction that every surviving conformally invariant field must match, one side to the other. However, this does not apply to the Hawking Points because almost all the information in a galactic cluster in a previous aeon becomes contained within a supermassive black hole whose ultimate fate in that aeon is to become the resulting Hawking radiation which is all squeezed into the Hawking Point in our aeon. However, the Hawking Points themselves will swallow almost all the degrees of freedom in the galactic clusters of the previous aeon. Each photon that apart from those that end up in a Hawking Point goes exactly to a corresponding photon, as it traverses the 3-surface joining one aeon to the next. However, a large majority of the surviving photons in a galactic cluster go into the Hawking Point.

This would appear to be a serious paradox, since we must suppose that, during the course of the evolution of each aeon, where the Second Law is expected to hold throughout, that the entropy should be enormously larger at the ultimate state of that aeon than it was at its big bang. The answer to this puzzle is a somewhat subtle matter. In observable physical processes, there is no violation of the Second Law, but to understand this we must return to the Boltzmann definition of entropy, as given in Section 3. We require knowing the volume V of the relevant coarse-graining region phase space appropriate to the state of the universe, and for that we need to know the appropriate phase space for the universe. I have already drawn the reader's attention, in Section 6, to the fact that, during most of the universe's evolution, by far the largest entropy increasing process is the formation of black holes. When a black hole grows through swallowing matter, the entropy continues to increase, but eventually, when the universe cools down and thins out, as it is expected eventually to do owing to its (accelerated) expansion, then as explained in Section 7 the temperature of the universe becomes cooler even than the tiny (Hawking) values that black holes have. At that point the hole shrinks in size as mass-energy is carried away in Hawking radiation. The process has the effect of continuing to be in accordance with the second law, but there is a subtlety here. According to Hawking's original analysis,[21] this process takes place because part of the information is swallowed by the black hole's singularity. Hawking argued that this process would continue until the eventual disappearance of the hole in a "pop" (an explosion in which the hole's final mass of only about 10^{-5}g is converted into energy), and that the information, swallowed by the hole's space-time singularity, must finally be considered to be lost to the universe.

In more recent years, many physicists (and eventually even Hawking himself) have tried to argue that the information is somehow restored through such effects as subtle correlations in the Hawking radiation. However it has long been my opinion, and has also been strongly argued by others,[22] that this is not correct and that the information is indeed lost. This applies, more importantly, also to the information in the Hawking radiation as the black hole evaporates away in a very remote future of that aeon. Irrespective of what happens at the black hole singularity this information is also lost in the Hawking Point since in the subsequent aeon (our aeon) all this information is squashed into a region much smaller than a Planck scale. To put the matter more correctly, there is a loss of phase space that is available to the universe, during the swallowing of information by the black hole, and the relevant degrees of information become completely inaccessible to the universe once the black hole has finally evaporated away in Hawking's "pop". This leads to a "thinning down" of the universe's total phase space, so that as measured on the scale of the universe as a whole, when a black hole disappears the same coarse-graining region as before, as far as the rest of the universe is concerned, will count as having a smaller volume (and indeed a smaller dimension) than it had before. In CCC, this thinning down of phase space volume eventually completely compensates for the increase in entropy that had occurred throughout the aeon's history, so that the hugely reduced degrees of freedom that are available in the aeon's final exponentially expanding state can match completely with those that are available at the big bang of the succeeding aeon!

---

21 Hawking (1975, 1976a, b).

22 See Braunstein and Pati (2007).

Clearly CCC raises many issues here which need detailed checking for their internal consistency. And there are many others which will have to stand up to the rigours of observation and future experiment. It will be of great interest to see how all these matters develop in the future.

References

An, D., K.A. Meissner, and P. Nurowski. 2018. "Ring Type Structures in the Planck map of the CMB". *Monthly Notices of the Roy. Ast. Soc.* 473(3): 3251–3255.

An, D., K.A. Meissner, P. Nurowski, and R. Penrose. 2020. "Apparent Evidence for Hawking Points in the CMB Sky". *Monthly Notices of the Roy. Ast. Soc.* 475(3): 3403–3408.

Bjorken, J., and S. Drell. 1965. *Relativistic Quantum Fields*. New York: McGraw-Hill.

Braunstein, S.L., and A.K. Pati. 2007. "Quantum Information Cannot Be Completely Hidden in Correlations: Implications for the Black-Hole Information Paradox". *Phys. Rev. Lett.* 98: 080502.

Friedrich, H. 1986. "On the Existence of *n*-Geodesically Complete or Future Complete Solutions of Einstein's Field Equations with Smooth Asymptotic Structure". *Commun. Math. Phys.* 107: 587–609.

Gurzadyan, V.G., and R. Penrose. 2013. "On CCC-Predicted Concentric Low-Variance Circles in the CMB sky". *Eur. Phys. J. Plus* 128: 22. 1–17. https://doi.org/10.1140/epjp/i2013-13022-4.

Guth, A.H. 1981. "Inflationary Universe: A Possible Solution to the Horizon and Flatness Problems". *Phys. Rev.* D 23: 347–356.

Hawking, S.W. 1975. "Particle Creation by Black Holes". *Comm. Math. Phys.* 43: 199–220.

Hawking, S.W. 1976a. "Black Holes and Thermodynamics". *Phys. Rev.* D 13 (2): 191–197.

Hawking, S.W. 1976b. "Breakdown of Predictability in Gravitational Collapse". *Phys. Rev.* D 14: 2460–2473.

Kronheimer E.H., and R. Penrose. 1967. "On the Structure of Causal Spaces". *Proc. Camb. Phil Soc.* 63: 481–501.

Meissner, K.A., P. Nurowski, and B. Ruszczycki. 2013. "Structures in the Microwave Background Radiation". *Proc. R. Soc.* A469: 2155, 20130116. arXiv: 1207.2498[astro-ph.CO].

Misner, C.W. 1968. "The Isotropy of the Universe". *Astrophys. J.* 151: 431–457.

Penrose, R. 1965. "Zero Rest-Mass Fields Including Gravitation: Asymptotic Behaviour". *Proc. Roy. Soc. London* A284: 159–203.

Penrose, R. 1979. "Singularities and Time-Asymmetry". In *General Relativity: An Einstein Centenary*, edited by S.W. Hawking and W. Israel. Cambridge: Cambridge University Press.

Penrose, R. 1990. "Difficulties with Inflationary Cosmology". In *Proceedings of the 14th Texas Symposium on Relativistic Astrophysics*, edited by E. Fenves. *N.Y. Acad. Sci.*, New York.

Penrose, R. 2004. *The Road to Reality: A Complete Guide to the Laws of the Universe*. London: Jonathan Cape.

Penrose, R. 2005. http://www.newton.cam.ac.uk/webseminars/pg+ws/2005/gmr/gmrw04/1107/penrose/.

Perlmutter, S. et al. 1998. "Cosmology from Type Ia Supernovae". *Bull. Am. Astron.* 29 [astro-ph/9812473].

Rindler, W. 2001. *Relativity: Special, General, and Cosmological*. Oxford: Oxford University Press.

Rugh, S.E., and H. Zinkernagel. 2007. "Cosmology and the Meaning of Time". Symposium, "The Socrates Spirit", Section for Philosophy and the Foundations of Physics, Hellabaekgade 27, Copenhagen N, Denmark.
Steinhardt, P.J., and N. Turok. 2002. "A Cyclic Model of the Universe". *Science* 296, 1436–1439. doi:10.1126/science.1070462 [arXiv:hep-th/0111030].
Steinhardt, P.J., and N. Turok. 2007. *Endless Universe: Beyond the Big Bang*. New York: Doubleday.
Tod K.P. 2003. "Isotropic Cosmological Singularities: Other Matter Models". *Class. Quantum Grav.* 20: 521–534. doi:10.1088/0264-9381/20/3/309 [arXiv:gr-qc/0209071].
Veneziano, G. 1998. "A Simple/Short Introduction to Pre-Big-Bang Physics/Cosmology". [arXiv:hep-th/9802057].

Gerard 't Hooft

How Studying Black Hole Theory May Help Us to Quantize Gravity

Abstract: Black holes are more than just odd-looking curiosities in gravity theory. They uniquely intertwine the basic principles of General Relativity with those of Quantum Theory. Just by demanding that they nevertheless obey acceptable laws of dynamics, just like stars and planets, we hit upon strange structures that must play key roles in the quantum effects that we expect in the gravitational force at ultrashort distance scales.

It is explained why, in our approach to address the problem of information conservation, the usual expression for the temperature of Hawking's radiation is off by a factor 2.

1 Introduction

Gravitation is arguably the most elementary force in physics. It appears to be directly linked to a fundamental principle:

Invariance of physical laws under general coordinate transformations in space and time.

As is well-known, other forces among the elementary particles of matter are based on very similar symmetry principles, notably those of local gauge theories, but the symmetry principle that governs the gravitational force seems to be at the basis of all symmetries in nature, that of general transformations among all coordinates for space and time. And more remarkably, combining this symmetry principle with quantum mechanics seems to lead to novel and unique clashes with what we thought we knew about forces and dynamics.

It was thought that history of science had given us a significant clue: *"We will get this thing done within half a century or so!"*

Unfortunately, in spite of a tremendous amount of work and numerous essential, novel pieces of insight, we still are confronted with mysteries at this point. What are we doing wrong?

Acknowledgments: We thank N. Sanchez, N. Gaddam, F. Feleppa, N. Groenenboom, and S. Kumar for discussions.

https://doi.org/10.1515/9783111313610-008

Most conventional theories, focusing either on physics at the Planck scale (some 10^{-33} cm or 10^{-44} seconds), or on the scale of cosmological theories (ranging up to the size of the universe), assume that the quantum formalism is mandatory:

Start with the existence of a Hilbert space, formulate a law for computing quantum amplitudes, and assume that all physical phenomena covered by the theory can be described in terms of these amplitudes, *even if* there is no need to agree on what it is that these amplitudes actually describe. According to many researchers, discussions on the foundations of quantum mechanics itself have come to a dead end. Here however, we emphasize that quantum amplitudes can actually be reduced to being a *vector representation* of phenomena based on extremely mundane forms of logic ('t Hooft 2016). One may ignore the usual conundrums of the 'collapse of the wave function', the role played by 'pilot waves', and even the existence of uncountably many distinct universes (the 'many world' hypothesis). And indeed we must worry about the nature of general coordinate transformations that we thought to have under control.

A deeper study of black holes (Schwarzschild 1916), and their relation with the laws of quantum mechanics (Hawking 1975), may teach us new and very important things. Black holes seem to be just the most basic solutions of Einstein's field equations, but it is not automatically guaranteed that their connection with quantum mechanics will be anything ordinary. In our telescopes it seems that black holes may be just a special kind of burned-out stars. But if conventional theories for the quantum treatment are not completely wrong, one does not get 'ordinary behavior' from our equations. Something is wrong. So, we advertise a closer look at those laws. Just by postulating that everything hangs together in a very orderly manner, this may force us to rephrase those equations in a way that might become very revealing.

2 Schwarzschild and the Tortoise Metric

Much of the material described in this section and the next one has appeared in several previous articles by this author ('t Hooft 1985; 1987; 2022). Readers familiar with this can just briefly scan this part of the paper, but we shall refer to it when we continue.

Shortly after Einstein published his theory of General Relativity, the astronomer Karl Schwarzschild (Schwarzschild 1916) realized that assuming spherical symmetry enables us to write down an exact solution of these equations. The metric one arrives at, in modern notation, reads

$$ds^2 = -dt^2\left(1-\frac{2GM}{r}\right)+\left(1-\frac{2GM}{r}\right)^{-1}dr^2+r^2d\Omega^2, \tag{1}$$

where r represents distance from the origin, t is a time coordinate, and Ω stands for the solid angle coordinates (θ,ϕ). The variables dr and dt stand for infinitesimal differences for the coordinates of two adjacent points in space-time, and $d\Omega$ stands short for their infinitesimal angular separation, $d\Omega^2 = d\theta^2 + \sin^2\theta\, d\phi^2$. M is a parameter that stands for the total mass as would be perceived by an observer at infinity (see Fig. 1a).

This metric is independent of time t, and we see that the contributions from dr and dt, as given in the first two terms in Eq. (1) both switch signs when r crosses the point $r = 2GM$. Here, G is Newton's gravitational constant, and the freely adjustable parameter M represents the mass of this object. Much later, it would become known as a 'black hole'.

The apparent singularity at $r = 2GM$ is not real, it is a coordinate artifact. One can replace the longitudinal coordinates r and t by a new set, x and y (see Fig. 1b) as follows:

$$xy=\left(\frac{r}{2GM}-1\right)\,e^{r/2GM}; \tag{2}$$

$$\frac{y}{x}=e^{t/2GM}. \tag{3}$$

These coordinates are called after Kruskal (Kruskal 1960) and Szekeres (Szekeres 1960), and shall be referred to here as 'tortoise coordinates' for short, just because they replace the infinitely slow geodesics to and from the horizon, by ordinary geodesics crossing a light front.

In these coordinates, the singularities at $r \to 2GM$ disappear:

$$ds^2=\frac{32(GM)^3}{r}e^{-r/GM}\,dxdy+r^2d\Omega^2. \tag{4}$$

However, here we discover that there are two horizons, not one: the *future event horizon*, at $x = 0$, to which all absorbed matter particles move, and the *past event horizon*, at $y = 0$, which may emit particles.

The tortoise coordinates show that the Schwarzschild space-time has a natural extension, from region *I*, where $x > 0$ and $y > 0$, to regions *II*, *III*, and *IV*. Region *I* is the physical region. Here, we can draw the geodesics for all in-going particles, and all out-going particles (see Fig. 2).

Believing that black holes should not be approached as magic wands, but should be understood as normal physical objects that can absorb things and emit things, leads one to believe that in-particles (short for particles going into the

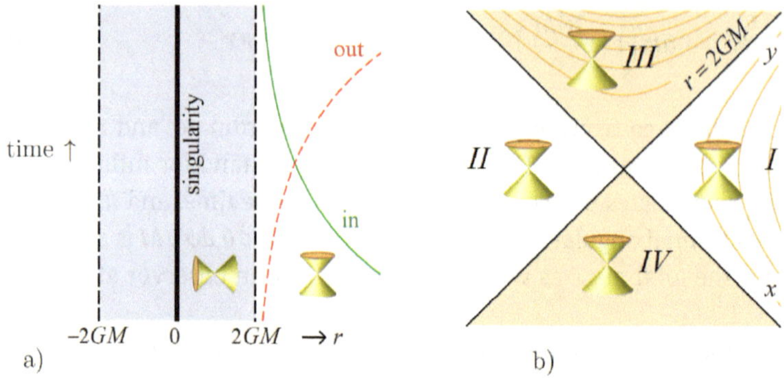

Fig. 1: a: The Schwarzschild metric. Vertical dashed lines: the event horizon. Curved lines: light like radial geodesics: y = const (smooth), and x = const (dashed). Cones: the orientation of the local light cones. Angular coordinates ($\theta;\varphi$) are not shown. b: The Tortoise coordinates (x; y), showing regions *I* - *IV* and the orientations of the local light cones. Curves show r = const. lines.

black hole) transmit all information they contain towards the out-particles. At first sight, this seems to violate causality, since the points where in-particles cross the future event horizon lie in the far future while the out-particles emerge in the far past (see Fig. 2). However, they meet each other half way, and if this point lies much further away than a Planck length from the horizon crossing point (point 0 in the figure), then the contradiction becomes insignificant. This separation decreases exponentially with the time span lying between the in-orbit and the out-orbit, so the Planck domain is reached quite quickly.

But more to the point, we can compute what happens here. It is the *gravitational force between in-particles and out-particles that does the job of information transfer* (as soon as their longitudinal coordinates meet), and, it acts locally as a sonic boom; there is no contradiction with causality at all.

3 The Shapiro Effect

How, by gravity, an in-particle 1 affects an out-particle 2 – and vice versa – is an elementary exercise in General Relativity. Consider first a flat background. One can take particle 1 to be at rest, and consider its metric as a Schwarzschild metric. Then let this particle move with a velocity close to c. Suppose now another light particle, 2, moves in the opposite direction. It experiences the gravitational field of particle 1 as being highly compressed by the Lorentz transformation. Immediately before, and immediately after the encounter, space-time is experienced as being flat (in the home frame of particle 1, the particles are then far apart). So the

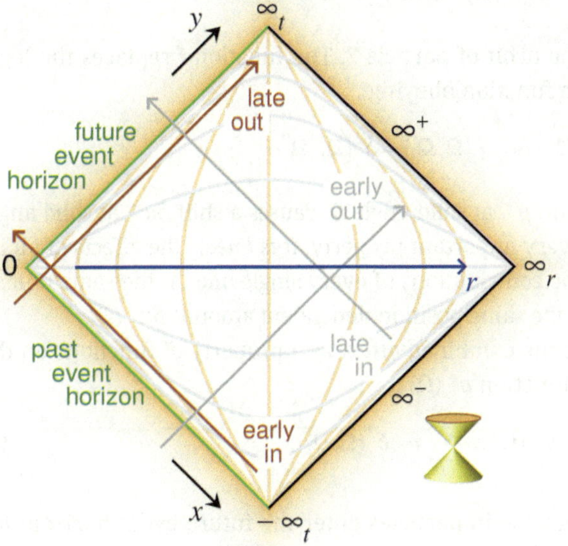

Fig. 2: Region I, the physical region (see Fig. 1, b), showing early and late particles going in and out. Everywhere in this region the light cones are oriented as the one shown.

gravitational field of particle 1 is experienced as a sonic boom, comparable to Cherenkov radiation (Cerenkov 1937).

During this encounter, what happens is closely related to the Shapiro effect (Shapiro 1964) when a light ray grazes past the Sun or another heavy body. The geodesic of the light ray is delayed. An elementary calculation (Aichelburg and Sexl 1971) shows that this amounts to a shift δu^μ of particle 2 in the light cone direction of the motion of particle 1. If particle 1 has a light cone momentum p^μ then one finds

$$\delta u^\mu = -4Gp^\mu \log|\tilde{x} - \tilde{x}'|. \tag{5}$$

Here, $|\tilde{x} - \tilde{x}'|$.is the transverse separation at the closest encounter (impact parameter).

This was the effect in a flat background space-time. Now consider particles moving in and out of a black hole (Dray and 't Hooft 1985), entirely in the longitudinal direction (close to the horizons, the transverse motion, and shift, are neglected for the time being). For a particle 1 moving in the longitudinal p^- direction, this shift can be written as ('t Hooft 1985; 1987)

$$\delta u^-(\Omega) = 8\pi G\, f(\Omega, \Omega'), \tag{6}$$

where δu^- is the shift in the orbit of particle 2. The function f replaces the logarithm in Eq. (5). It is a Green function obeying

$$(1 - \Delta_\Omega)\, f(\Omega, \Omega') = \delta^2(\Omega,\ \Omega'). \tag{7}$$

An in-particle with momentum p^- at solid angle Ω' causes a shift δu^- at solid angle Ω. This Shapiro effect has a very important property: *it is linear*: the effects of all in-particles is just the sum of the contributions of every single one. At the same time, it displaces all out-particles by the same position-dependent amount $\delta u^-(\Omega)$.

This enables us to write the combined effect as a property of functions on the angles Ω, Ω'. Write a single function $p^-(\Omega)$ as

$$p^-(\Omega) \equiv \sum_i p_i^-\, \delta^2(\Omega, \Omega_i), \tag{8}$$

where Ω_i are the spots where the in-particles enter the future event horizon; for all out-particles we write:

$$u^-(\Omega_i) = u_i^-. \tag{9}$$

The Shapiro effect can now be written as follows: if we add an in-particle with momentum δp_i^- at solid angle Ω_i then it modifies the position $u^-(\Omega)$ of all out-particles, in accordance with the equations

$$u^-(\Omega) = 8\pi G \int d^2\Omega' f(\Omega, \Omega') p^-(\Omega'), \tag{10}$$

where the Green function $f(\Omega, \Omega')$ obeys (7). We can write

$$(1 - \Delta_\Omega)\, u^-(\Omega) = 8\pi G\, p^-(\Omega). \tag{11}$$

Notice that we replaced the shift equation (6) for one particle going in and one particle going out, by a single functional equation (10) for all particles in and all particles out. The only difference this makes is that we could omit the symbol δ that indicates *change*. Before and after any change, out-positions relate to in-momenta in the same way. The equation suggests that the origin, where all positions are zero and no particles are going in at all, should be chosen as the origin of this space.

4 Quantum Mechanics

In the previous section, we encountered fundamental new equations relating positions and momenta of in- and out-going particles. This begs for an investigation as to what happens when we introduce quantum mechanics. Close to the origin,[1] all particles obey

$$\left[u_i^{\pm}, p_j^{\mp}\right] = i\delta_{ij}, \quad [u^{\pm}, p^{\pm}] = 0, \tag{12}$$

so that, for the functions $u^{\pm}(\Omega)$ and $p^{\pm}(\Omega)$, we can write

$$[u^{\pm}(\Omega),\, p^{\mp}(\Omega')] = i\delta^2(\Omega, \Omega'), \tag{13}$$

$$[u^{\pm}(\Omega),\, p^{\pm}(\Omega')] = 0. \tag{14}$$

In combination with Eqs. (10) and (11), these equations become very powerful. The equations are local now, and, most important, they are linear. Linearity and locality might get broken when we refine these equations for large transverse momenta, but we believe that the section for sufficiently low transverse momenta can be kept by itself as the dominant contribution to the information processing mechanism. Due to linearity, it is meaningful to expand the functions $u^{\pm}$ and $p^{\pm}$ in spherical harmonics:

$$u^{\pm}(\Omega) \equiv \sum_{\ell,m} u^{\pm}_{\ell m} Y_{\ell m}(\Omega), \; p^{\pm}(\Omega) \equiv \sum_{\ell,m} p^{\pm}_{\ell m} Y_{\ell m}(\Omega), \tag{15}$$

where the sum goes over all nonnegative integer values of ℓ, and $-\ell \le m \le \ell$.

For the spherical harmonics, the operator Δ diagonalizes into $-\ell(\ell+1)$, so that we can write

$$u^{-}_{out,\ell m} = \frac{8\pi G}{\ell^2+\ell+1}\, p^{-}_{in,\ell m}, \; u^{+}_{in,\ell m} = -\frac{8\pi G}{\ell^2+\ell+1}\, p^{+}_{out,\ell m}, \tag{16}$$

where we wrote the subscripts 'in' and 'out', to indicate that these operators refer to in- or out-particles. The minus sign in the last equation is understood as a consequence of the antisymmetry of the commutator while no minus signs had been inserted in Eqs. (12):

---

1 The process of information transfer, in which we are now interested, takes place at Planckian distance scales. As long as the Schwarzschild metric parameters, typically in the order of *GM*, are large compared to this, our analysis applies.

$$\left[u^+_{\ell m}, u^-_{\ell' m'}\right] = \frac{8\pi i G}{\ell^2+\ell+1}\delta_{\ell\ell'}\delta_{mm'}. \tag{17}$$

Important remarks:

- There is only one real independent dynamical variable u^+, and a variable p^- canonically associated to it, at every value of the pair of integers (ℓ, m). So the variables in Eqs. (12) are replaced by a one-dimensional quantum mechanical pair at each ℓ and m.
- Different (ℓ, m) values all commute. Therefore, in the approximations used, the complete set of all quantum states will be just the product of all quantum states $|u^+\rangle_{\ell m}$ at all (ℓ, m). By taking their one-dimensional Fourier transforms, one gets the states $|p^-\rangle_{\ell m}$ instead.

Thus, the out-particles emerge as being the Fourier transforms of the in-particles, a very simple algebra at this stage.

Note that, replacing one single species of matter by a set of multiple matter species, would lead to contradictions (there exists only one type of gravitational force). There exists only one form of matter which we subject to the rules (12), and therefore we cannot apply second quantization in the longitudinal direction; only the Ω dependence can be viewed as a reduced, two-dimensional second quantization process. Having single particle states only for the $u^\pm$ dependence, but second quantization only in the θ,ϕ direction could be called '1.5th quantization'.

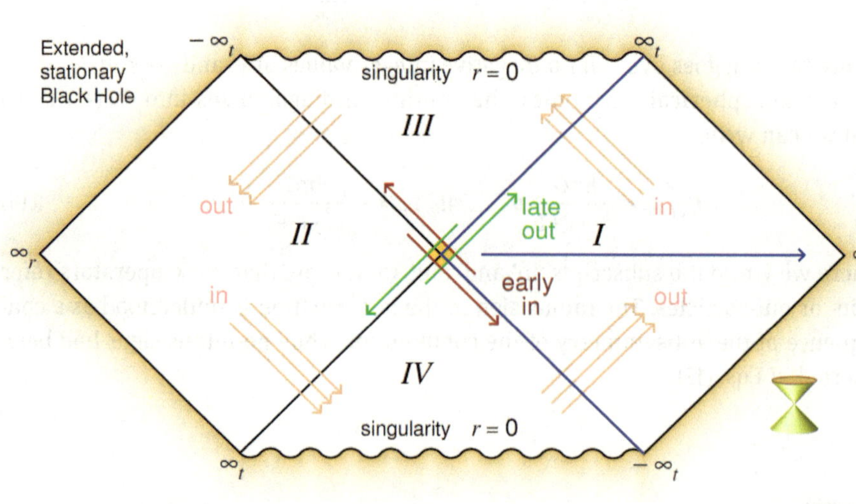

Fig. 3: Penrose diagram generated from the tortoise coordinates *x* and *y*, by mapping these both on a compact segment [1,1]. See text.

Everything seems to come together, but there are still some hurdles to take. By mapping the Kruskal-Szekeres coordinates onto compact segments, one gets the Penrose diagram for an eternal black hole (see Fig. 3). It was shown above that the out-particles are obtained from the in-particles by Fourier transforming the wave functions, call these $\langle u^+|\psi\rangle_{\ell,m}$. The Fourier transformation is a unitary process, so one might think that we obtained the complete black hole evolution operator *U(t)*. The fact that, so far, we concentrated only on *eternal* black holes, is not a problem. At time scales longer than *O*(*M* log*M*) in Planck units, the wave functions get completely scrambled by the Shapiro process, and the component of Hilbert space that would represent either collapse, or a final evaporation event, is expected to be small compared to the one spanned by the states that we do take into account.

A problem that we still have, at this stage, is that there are two asymptotic regions, *I* and *II*. Physically, we need unitarity only for one single domain, region *I* itself. But by generating the shifts δu, the Shapiro process will transport states right across the horizon, and when we attempt to calculate the combined effects, we find that unitarity is violated by that process: particle wave functions in region *I* will generate Fourier transforms that cover both regions *I* and *II*.

It is true that, most of this process takes place in a Planckian domain close to the origin as indicated in Fig. 3, but, as time goes on, local Lorentz boosts will quickly send these out-states to infinity, spreading themselves both over regions *I* and *II*.

In the past, we proposed to cure this problem by assuming region *II* to correspond to the *antipodes* (in terms of the angles θ and ϕ) of region *I*. But this did not work as assumed. In terms of the harmonic functions $Y_{\ell m}(\Omega)$, going to the antipodes means multiplying the angular wave functions by $(-1)^\ell$. For odd ℓ, this actually cancels the minus sign transporting us from region *I* to region *II* or back, causing our calculations to fail.

We now believe that region *II* is an *exact quantum copy* ('t Hooft 2022) of all states in region *I*. This means that we would limit ourselves to wave functions that are *even* under the interchange $I \leftrightarrow II$. This could also mean that all information in region *I* is accurately copied into region *II*, and this remains to be the case for the Fourier transforms (the Fourier transform of an even function is even as well).

This brings us very close to an ancient proposal by the author ('t Hooft 1984), to identify the states in region *II* as the set of bra states associated to the kets in region *I*. At first sight this still may generate problems, since energy reversal from *I* to *II* requires us to take real functions to be even but imaginary functions to be odd. This property however, would not be passed on correctly by the Fourier transforms. We suspect that this problem can be solved by restricting ourselves to *real* wave functions only.

Limiting ourselves to real wave functions only, now leads to serious questions concerning the very foundations of quantum mechanics. The picture we obtained so far is that there are three representations for the wave functions of the in- and out-particles: first, we have the original Schwarzschild metric, and all in- and outgoing particles can be represented as wave functions depending on r, t, θ, and ϕ. We subsequently write the same metric in terms of the tortoise coordinates x, y, θ, and ϕ. Using the same wave functions, this gives us the quantum states in region I when x and y are positive. Finally, we have the negative values of x and y. Again, the same wave functions now represent what goes on in region II.

The equations for these wave functions in (x, y) space will be as usual. The problems just mentioned above concern the time reversal in region II. It is more than just an administrative difficulty. It seems that the concepts *lowest energy state* and *highest energy state* are interchanged there. If this is allowable then we can treat the entire system in terms of a single set of quantum field theoretical equations, where the technical difficulties in gluing the positive and negative values of x and y together must be more carefully addressed than we were able to do here.

5 On the Hawking Temperature

An issue not yet addressed in detail is the value of Hawking’s temperature,

$$kT_{Hawking} = \frac{1}{8\pi G M_{BH}}. \tag{18}$$

It is connected to the calculation of the Hawking entropy,

$$\beta = \frac{1}{kT} = \frac{\partial S}{\partial E} = 8\pi G M. \tag{19}$$

Now these values represent the contributions of the quantum states both in regions I and II. If, however, region II is nothing but a quantum clone of region I, then it should not contribute to the entropy at all, so that only half of the entropy is left:

$$S = \frac{1}{2}\ 4\pi G\, M^2; \beta = \frac{1}{2}\ \beta_{Hawking}; kT = 2kT_{Hawking}. \tag{20}$$

This happens to be the temperature we anticipated in our treatment of Ref. (’t Hooft 1984) using different arguments: the picture of regions I and II together as representing bras and kets, turns our expressions into a density matrix. To arrive at expectation values one should not take the absolute square of a density matrix element but instead, trace it with the relevant operators. Replacing the quadratic expressions by linear ones reduces the entropy by a factor 2.

Now both arguments mentioned above, to indicate that the value of the temperature is affected by a factor 2, seem to be mostly hand-waving. Author warns Reader against believing that N hand-waving arguments may add up to something more believable than a single one. But we have a more powerful argument, using the periodicity of the evolution operator in the Euclidean time direction. This was the calculation first employed by Hawking (Hawking 1975) and others.

Consider an outside observer A, and an observer B in the effectively flat space-time at the origin of the Penrose diagram (Figs. 1b and 3). If the outside observer A performs a time translation by a constant amount,

$$U(\delta t) = e^{-i\,\delta t\,H}, \tag{21}$$

then the tortoise variable x in Eqs. (2), (3) is multiplied by $e^{\delta t/4GM}$ and y is divided by the same amount. For observer B that is a Lorentz transformation at the origin. Now take the same transformation but replace δt by an imaginary number $-i\beta$. This is a rotation:

$$x \rightarrow x e^{-i\beta/4GM}, \; y \rightarrow y e^{i\beta/4GM}. \tag{22}$$

Writing

$$x = \varrho e^{i\varphi}, \; y = \varrho e^{-i\varphi}, \tag{23}$$

one gets a real, Euclidean space-time in polar coordinates (by using the coordinates $(x+y)$ and $-i(x-y)$). The standard calculation, computing the free energy F from $Tr(e^{-\beta H})$, now assumes that only full rotations, $\phi \rightarrow \phi + 2\pi n$ lead space-time back to its original orientation. The Green functions to be considered are therefore periodic with period 2π in ϕ. This is how β takes the value $8\pi GM$. The identification of the Euclidean periodicity β with $1/kT$ is well-known in condensed matter physics and in lattice theories for elementary particles. Thus one gets Eqs. (18) and (19) for the Hawking temperature.

Now in our theory, we have that region II has x and y replaced by $-x$ and $-y$.

There, it is a quantum clone of region I. All our Green functions must be the same in region II and in region I. Thus, our Green functions must return to their original values when ϕ is replace by $\phi + \pi$. This gives us the temperature (20).

6 Conclusion

In some sense, our theory of black holes is extremely conservative. In our treatment there is no place for 'entanglement' issues, and not even for attaching significant roles for the 'Page time' (Page 1993). Page asks for explicit attention to very large time scales, where a significant fraction of the initial entropy is carried away by the radiated particles. But as soon as the radiated particles have moved further than a few Planck lengths away from the horizon crossing point, conventional laws of physics take over, entanglement or not. Any respectable theory of the production of these particles must include a chapter on how these particles are expected to make this transition to free or almost free particle states in the outside world.

In fact, this seems to be easy enough. The total mass M of a black hole together with the surrounding particles is strictly constant. But only the particles that are still – or already – close to their respective horizons, contribute to the mass in practice. So, as soon as we see particles far enough away from the black hole, we should modify our description in terms of a black hole with mass M', slightly different from M, by subtracting the contribution of the particles that are sufficiently far away, at any particular moment of time.

Note that, in our approximations, the masses of the radiated particles are small compared to that of the black hole itself, so that the time-dependent corrections, as described above, are infinitesimal in any case. I think we performed important ground work to formulate a precise framework for handling the physics of black holes. In ancient times when solitons, magnetic monopoles, field theoretical strings and membranes were regarded as novel predictions of gauge theories and related concepts were discovered so that their properties were calculated, no new physical principles had to be called upon. For black holes, including of course the rotating and charged versions, the situation is not at all that simple, and it could even be that the physical laws at small scales will have to be thoroughly revised in order to obtain a harmonic picture of what goes on. This would actually be welcome, since out present views of Planckian scale physics appear to lead to paradoxes: how should one handle energies beyond that scale, and how can we reconcile that with quantum mechanics?

The author's personal view is that both quantum mechanics and general coordinate invariance are aspects of matter and geometry that are calling for a more precise formulation of their foundations. It does not help to be 'agnostic', as some physicists declared, or to think that 'chaos' takes care of all contradictions.

Particularly, our results on Hawking's temperature are novel (they were already mentioned in 1984 ('t Hooft 1984) but hardly noticed by string theorists).

Some further thinking strongly suggests that, actually, any attempt to identify the information residing in region *II* with what is already present in region *I*, would reduce the Hawking entropy by a factor 2, and raise the temperature of the radiation accordingly.

Since conservation of information (in the sense of entropy in thermodynamics) is being considered more seriously these days, and in view of our considerably revised outcome (20) for temperature and entropy in black holes, we conclude that thorough revisions for quantum gravity itself may be opportune. Our proposals for real wave functions rather than complex ones, also suggests revisions for the foundation of quantum mechanics ('t Hooft 2021).

References

Aichelburg, Peter C., and Roman U. Sexl. 1971. "On the Gravitational Field of a Massless Particle". *General Relativity and Gravitation* 2: 303–312. http://dx.doi.org/10.1007/BF00758149.

Cerenkov, Pavel A. 1937. "Visible Radiation Produced by Electrons Moving in a Medium with Velocities Exceeding That of Light". *Physical Review* 52 (4): 378–379. http://dx.doi.org/10.1103/PhysRev.52.378.

Dray, Tevian, and Gerard 't Hooft. 1985. "The Gravitational Shock Wave of a Massless Particle". *Nuclear Physics B* 253(1): 173–188. https://doi.org/10.1016/0550-3213(85)90525-5.

Hawking, Stephen W. 1975. "Particle Creation by Black Holes". *Communication in Mathematical Physics* 43(3): 199–220. http://dx.doi.org/10.1007/BF02345020.

Kruskal, Martin D. 1960. "Maximal Extension of Schwarzschild Metric". *Physical Review* 119(5): 1743–1745. http://dx.doi.org/10.1103/PhysRev.119.1743.

Page, Don N. 1993. "Information in Black Hole Radiation". *Physical Review Letters* 71(23): 3743–3746. http://dx.doi.org/10.1103/PhysRevLett.71.3743.

Schwarzschild, Karl. 1916. "Über das Gravitationsfeld eines Massenpunktes nach der Einsteinschen Theorie". *Sitzungsberichte der Königlich Preussischen Akademie der Wissenschaften* 7: 189–196 ("On the Gravitational Field of a Mass Point according to Einstein's Theory", translated by S. Antoci and A. Loinger (1999). https://doi.org/10.48550/arXiv.physics/9905030.

Shapiro, Irwin I. 1964. "Fourth Test of General Relativity". *Physical Review Letters* 13: 789–791. https://doi.org/10.1103/PhysRevLett.13.789.

Szekeres, George 1960. "On the Singularities of a Riemannian Manifold". *Publ. Math. Debrecen* 7: 285–301 (now available at *General Relativity and Gravitation* 34 (2002). https://doi.org/10.1023/A:1020792830651).

't Hooft, Gerard. 1984. "Ambiguity of the Equivalence Principle and Hawking's Temperature". *Journal of Geometry and Physics* 1: 45–52. https://doi.org/10.1016/0393-0440(84)90013-5.

't Hooft, Gerard. 1985. "On the Quantum Structure of a Black Hole". *Nuclear Physics B* 256: 727–745. https://doi.org/10.1016/0550-3213(85)90418-3.

't Hooft, Gerard. 1987. "Strings from Gravity". *Physica Scripta* T 15: 143–150. http://dx.doi.org/10.1088/0031-8949/1987/T15/019.

’t Hooft, Gerard. 2016. *The Cellular Automaton Interpretation of Quantum Mechanics*. Berlin: Springer International.
’t Hooft, Gerard. 2021. “Explicit Construction of Local Hidden Variables for Any Quantum Theory up to Any Desired Accuracy”. arXiv:2103.04335 [quant-ph].
’t Hooft, Gerard. 2022. “Quantum Clones inside Black Holes”. *Universe* 8(10): 537. http://dx.doi.org/10.3390/universe8100537.

Fabio Scardigli

Uncertainty Principle and Gravity

Abstract: We give a pedagogical introduction to the Generalized Uncertainty Principle (GUP) by showing how it naturally emerges when the action of gravity is taken into account in measurement processes. Then we review some of the physical predictions of the GUP. We focus in particular on the bounds that present experimental tests can put on the value of the deformation parameter β, and on the prediction of black hole remnants, which are a good candidate for dark matter. In passing, we also quote a theoretical value computed for β, and swiftly comment on the vast parameter region still unexplored, and to be probed by future experiments.

1 Introduction

The uncertainty principle is the cornerstone of Quantum Mechanics (Heisenberg 1927). The principle, when referred to single point particles, can be formulated in a quite simple way:

It is impossible to measure, with an arbitrary precision, position and speed of a particle at the same instant in time.

The idea emerged when Heisenberg, in a seminal paper of 1927, discussed measurement processes in Quantum Theory. He illustrated the principle through a *gedanken experiment*, which from then is known under the name of "Heisenberg microscope argument" (Heisenberg 1930). Heisenberg's original idea was to measure position and momentum of a static particle, say an electron, by using a photon as a probe. The photon scatters off the electron, and by measuring the properties of the photon after the scattering, one would like to know the exact position x_e and momentum p_e of the electron at the instant of the scattering. However, since the photon has a wavelength λ, from the principles of wave optics follows that the uncertainty in the position of the electron is (at least) $\Delta x_e \simeq \lambda$. Moreover, the photon carries a momentum $p = E/c = h/\lambda$, which, during the scattering, is partially transferred to the electron in an *unknown* magnitude and direction. This implies that, just after the scattering, the uncertainty in the electron momentum amounts to (at most) $\Delta p_e \simeq p = h/\lambda$. Therefore, Heisenberg concluded that

$$\Delta x_e \, \Delta p_e \simeq \lambda \cdot \frac{h}{\lambda} \simeq h. \tag{1}$$

https://doi.org/10.1515/9783111313610-009

Successively, Robertson and Schrödinger (Robertson 1929; Schrödinger 1930) formulated the uncertainty principle for canonically conjugated variables, such as the position x and momentum p of a particle, in the form

$$\Delta x \, \Delta p \geq \frac{\hbar}{2}, \tag{2}$$

which is the expression commonly accepted today.

In those early approaches the gravitational interaction between particles was completely neglected, although this attitude was somehow justified by the huge weakness of gravity, when compared with other fundamental interactions. However, Heisenberg's heuristic approach paved the way to the formulation of Generalized Uncertainty Principle(s), GUPs (Bronstein 1936; Amati, Ciafaloni, and Veneziano 1987; Maggiore 1993; Kempf, Mangano, and Mann 1995; Scardigli 1999; Adler and Santiago 1999; Capozziello, Lambiase, and Scarpetta 2000; Scardigli and Casadio 2003), which originate just by properly taking into account the gravitational effects in the photon-particle interaction. For example, the unavoidable formation of (microscopic) black holes in the measurement process, or even the simple Newtonian photon-electron gravitational attraction, imply the existence of a *minimum testable length*, below which position measurements become meaningless.

2 Generalized Uncertainty Principle

A possible generalization of the classical Heisenberg argument goes roughly as follows (Scardigli 1999). According to Heisenberg principle, one might probe a length as small as one wishes, provided one uses enough energetic photons. In fact, the size of the observable detail is the same as the wavelength of the photon, $\delta x \simeq \lambda$. But a photon of wavelength λ brings an energy $E = hc/\lambda$, therefore

$$\delta x \simeq \frac{hc}{E}. \tag{3}$$

So that increasingly large energies allow to explore decreasingly small details. Incidentally, this is why we build particle accelerators ever larger and more powerful. If you want to probe small details, you need a lot of energy! Now, the interaction between a photon and, say, an electron, is punctual, as far as we know, namely the energy E carried by the photon is packed in a region of size at most δx, at the instant of the interaction. Even supposing that we ignore particle pair creation, however, gravity cannot be neglected. Here in fact gravity enters the game. In general, we know that any given mass M has an associated gravitational radius

$$R_g(M) = \frac{2GM}{c^2}, \tag{4}$$

or, in terms of energy, $R_g = 2GE/c^4$. The physical meaning of these equations is that, when a mass M, or an energy E, is packed inside its own gravitational radius R_g (E), then a black hole is formed. From the sphere of radius R_g (E), the so-called *event horizon*, not even light can escape, so that region appears black. Of course, inside a black hole one cannot watch, by definition! So, every detail which falls inside the sphere of radius R_g (E) is invisible for the outside observer. In terms of our measurement process, this means that, if the probing photon is too energetic, a micro black hole of size R_g (E) is formed, and this results into an unobservable region of size R_g (E). Essentially the event horizon screens smaller details in this region, and this results into a further error that should be added to the usual quantum mechanical error predicted by the standard Heisenberg principle. Once the micro black hole is formed, even if we try to increase the photon energy E to "see" more refined details (according to the equation $\delta x \simeq hc/E$), what we are really doing is just "to feed" the micro black hole. It will grow, and will hide more and more details behind its event horizon of radius R_g (E). The above considerations can be summarized in a formula. So, the smallest detail that a photon of energy E can actually distinguish is

$$\delta x \simeq \frac{hc}{E} + \beta\, R_g(E). \tag{5}$$

Here β is a dimensionless parameter, which is not in principle fixed by the theory, but it is generally assumed to be of the order of unity. An explicit analytic calculation of β in (Scardigli, Lambiase and Vagenas 2017) has confirmed the circumstance. However, many studies have appeared in the literature, with the aim to set experimental bounds on β (see, for instance, Scardigli and Casadio 2015). For a standard Schwarzschild black hole the gravitational radius is given by Eq. (4), so Eq. (5) now reads

$$\delta x \simeq \frac{hc}{E} + \beta\, \frac{2GE}{c^4}. \tag{6}$$

Since for photons the dispersion relation $E = pc$ holds, and repeating the considerations seen in Sec.1, we can say that the uncertainty in the electron's momentum just after the scattering is, at most, $\Delta p \simeq p = E/c$. So, introducing the definitions of Planck length, and Planck mass, respectively, as

$$\ell_p^2 = \frac{G\hbar}{c^3}, \quad m_p c \ell_p = \frac{\hbar}{2}, \tag{7}$$

we can write the new *Generalized Uncertainty Principle* (GUP) as

$$\Delta x\ \Delta p \geq \frac{\hbar}{2}\left[1+\beta\left(\frac{\Delta p}{m_p c}\right)^2\right]. \tag{8}$$

It is remarkable that many different approaches lead to the same formulation of the GUP. Not only *gedanken experiments* involving micro black holes, but also arguments considering large black holes (Maggiore 1993), or arguments from string theory (Amati, Ciafaloni, and Veneziano 1987), and even considerations involving just the simple Newtonian gravity (Adler and Santiago 1999), they all lead to the same Eq. (8). The diagram in Fig. 1 summarizes the situation described by Eqs. (6), (8).

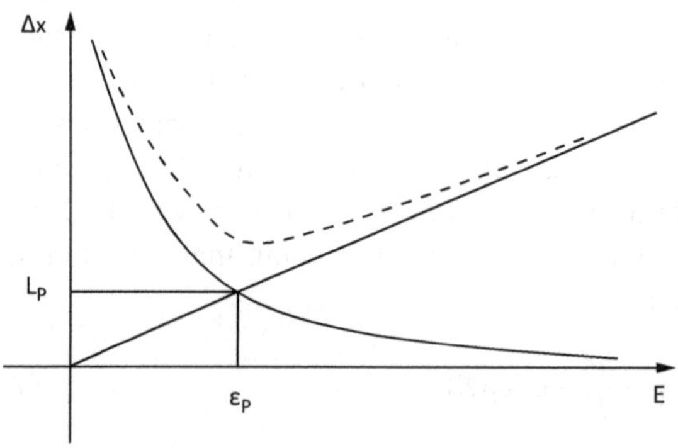

Fig. 1: Illustrations of Eqs. (6), (8).

If gravity did not exist at all, then one could probe lengths small at will, provided to use enough large energies. However, the very existence of gravity, and of a gravitational radius linearly proportional to the energy of the probe itself, modifies the Heisenberg hyperbole with a straight line, and the actual uncertainty $\Delta x(E)$ is described by the dashed line in Fig. 1. Thus, Eqs. (6), (8) imply the existence of a *minimum observable length* (or detail), which can be easily computed to be $L_{min} \simeq l_p\sqrt{\beta}$. Such small details can be observed, in principle, only by using probes (e.g. photons) of Planckian energies, namely $E \simeq E_p/\sqrt{\beta} = m_p c^2/\sqrt{\beta}$. To have an idea of how far still we are from reaching those energies, we just remind that the most powerful particle accelerator in the world, the LHC of CERN, in Geneva, reaches at its best 10^4 GeV, while the Planck energy is at $E_p \simeq 10^{19}$ GeV.

3 Deformed Quantum Mechanics: Tests of GUP

The generalized uncertainty relation (8) can be described also in terms of a fundamental commutator between the position $\hat{X}$ and momentum $\hat{P}$ operators, as

$$[\hat{X}, \hat{P}] = i\hbar \left(1 + \beta \frac{\hat{P}^2}{m_p^2 c^2}\right), \tag{9}$$

where the fundamental variables $\hat{X}, \hat{P}$ are thought to be high-energy operators useful, in particular, at or near the Planck scale. They have non-linear representations, $\hat{X}(\hat{x}),\ \hat{P}(\hat{p})$, in terms of the operators $\hat{x},\ \hat{p}$ which are the usual position and momentum operators at low energy, obeying the standard Heisenberg canonical commutator $[\hat{x},\ \hat{p}] = i\hbar$. Several authors (Kempf, Mangano, and Mann 1995; Brau 1999; Das and Vagenas 2008) have used the above approach to construct a *Deformed (or Generalized) Quantum Mechanics*. In this way it becomes possible to re-compute virtually any quantum phenomenon by using the GUP commutator (9). The results (each one obviously dependent on the deformation parameter β) have been compared with the experimental data, so to extract (upper) bounds on the parameter β. Besides, since by definition $[\hat{x},\ \hat{p}] \neq [\hat{X}, \hat{P}]$, the transformation $\hat{X}(\hat{x}),\ \hat{P}(\hat{p})$, is *non-canonical*, namely the correspondent Poisson brackets are not preserved. Therefore, the GUP is really describing new physics, doesn't simply describe the usual physics with different variables. A particularly useful example of such non-linear representations is the following. Consider (with $\beta_0 = \beta/m_p^2c^2$)

$$\begin{cases} \hat{X} = \hat{x} \\ \hat{P} = \hat{P}(\hat{p}) \end{cases} \text{and} \begin{cases} [\hat{X}, \hat{P}] = i\hbar \left(1 + \beta_0 \hat{P}^2\right) \\ [\hat{x}, \hat{p}] = i\hbar \end{cases}; \text{ then it follows } \hat{P} = \frac{1}{\sqrt{\beta_0}} \tan\left(\sqrt{\beta_0}\, \hat{p}\right).$$

If we use the standard representation of the low energy momentum $\hat{p} = -i\hbar\partial_x$, then the Hamiltonian written with the high-energy variables is actually equivalent to a low energy canonical Hamiltonian containing an infinite derivative kinetic term

$$\hat{H} = \frac{\hat{P}^2}{2m} + U(\hat{X}) = \frac{1}{2m\beta_0} \left[\tan\left(-i\hbar\sqrt{\beta_0}\, \partial_x\right)\right]^2 + U(\hat{x}). \tag{10}$$

Of course, the corrections, at the first order in β, to the standard quantum mechanical results are computed in perturbation theory by expanding the tangent function in (10). So the Hamiltonian governing a GUP-deformed quantum system will be (at the first order in β)

$$\hat{H} = \frac{\hat{P}^2}{2m} + U(\hat{X}) = \frac{\hat{p}^2}{2m} + U(\hat{x}) + \beta_0 \frac{\hat{p}^4}{3m}. \tag{11}$$

Using the above Hamiltonian, various different systems have been re-computed, and, by comparison with the experimental data, the following upper bounds for the deformation parameter β have been obtained (Tab. 1).

Tab. 1: Upper bounds on β.

^{87}Rb cold-atom-recoil experiment	$\lvert\beta\rvert \leq 10^{39}$
Hydrogen transition 1S-2S; Charmonium levels	$\lvert\beta\rvert \leq 10^{34}$
Electroweak scale	$\lvert\beta\rvert \leq 10^{34}$
Lamb shift	$\lvert\beta\rvert \leq 10^{20}$
Micro & Nano-mechanical oscillators (with masses ~ m_p)	$\lvert\beta\rvert \leq 10^{12}$
Sapphire mechanical resonator	$\lvert\beta\rvert \leq 10^{6}$

Such large upper bounds should not impress the reader. Somehow, we were to expect them, given the extremely large suppression of the deforming term in β, due to the smallness of the coefficient $1/m_p^2c^2 = 4l_p^2/\hbar^2$. As we said in the previous section, an exact theoretical calculation of β has been performed in (Scardigli, Lambiase, and Vagenas 2017), and it was found $\beta = 82\pi/5$, namely β is of order 10 (as already suspected on different theoretical grounds). A simple glimpse to the experimental bounds reported in Tab. 1 immediately testifies of the large (or perhaps huge) gap that experimental investigations still have to cover in order to arrive to test the theoretical value.

4 Hawking Temperature, GUP, and Black Hole Remnants

Among the many things that can be recomputed through the GUP there is the Hawking temperature of an evaporating black hole. Let us first derive the standard Hawking formula directly from the usual Heisenberg principle (Scardigli 1995). The argument proceeds as follows: suppose you have a bunch of Hawking photons just outside the event horizon of a Schwarzschild black hole of mass M. Further, suppose the photons are in a semiclassical regime, namely their wavelength is large, so that the equipartition theorem holds, and the average energy of a Hawking photon is given by $E = k_B T$. If the photons are close to the event horizon, then the uncertainty in their position is of the order of the geometrical size of the horizon, namely $\simeq 2\mu R_g$, where μ is a parameter that will be adjusted in order to match the actual Hawking temperature. Then, since relation (3) between δx and E should hold, we can also write

$$2\mu R_g = \frac{hc}{k_B T},$$

or

$$T = \frac{hc}{2\mu k_B R_g} = \frac{hc^3}{4\mu\ k_B GM} = \frac{\hbar c^3}{8\pi k_B GM}, \tag{12}$$

where the last equality holds if we choose $\mu = 4\pi^2$, and of course $\hbar = h/2\pi$. The above derivation uses the standard Heisenberg relation (3). But we can now repeat exactly the above steps by using instead the GUP relation (6). Then one ends up with a deformed Hawking temperature T satisfying the equation

$$M = \frac{\hbar c^3}{8\pi k_B GT} + \beta\, \frac{k_B T}{8\pi^2 c^2}. \tag{13}$$

Relation (13) is directly derived from the GUP, and of course reduces to standard Hawking formula for $\beta \to 0$. Relation (13) can be easily inverted, and then expanded to the first order in β, so obtaining

$$T = \frac{\hbar c^3}{8\pi k_B GM}\left(1 + \frac{\beta}{4\pi^2}\,\frac{m_p^2}{M^2} + \ldots\right). \tag{14}$$

The physical predictions of relation (13) are striking in many different sectors. Astrophysics and cosmology, among others, are fields deeply affected by the consequences of Eq. (13). Consider for example the diagram depicted in Fig. 2.

The green (bottom) line represents the standard mass/temperature relation (12) of Hawking for a Schwarzschild black hole, while the blue (top) line is the GUP-deformed mass/temperature relation (13). As we see, the usual Hawking formula predicts that the black hole evaporation goes on until a final *zero* mass and an *infinite* final temperature. On the contrary, according to the GUP-deformed formula, Eq. (13), the mass/temperature diagram has a minimum. So the evaporation starts with a large mass M, and halts when M reaches a minimum mass $M_{min} \sim m_p$ and a correspondent maximum *finite* temperature $T_{max} \sim T_p$ (being T_p the Planck temperature, defined as $k_B T_p = E_p$). The use of the GUP has the great advantage of avoiding an unphysical infinite final temperature. Actually, a more refined analysis reveals that the temperature drops to zero in the very last phases of the evaporation. So, GUP predicts the formation of small inert *remnants* as final product of the black hole's evaporation (Adler, Chen, Santiago 2001). These remnants do not

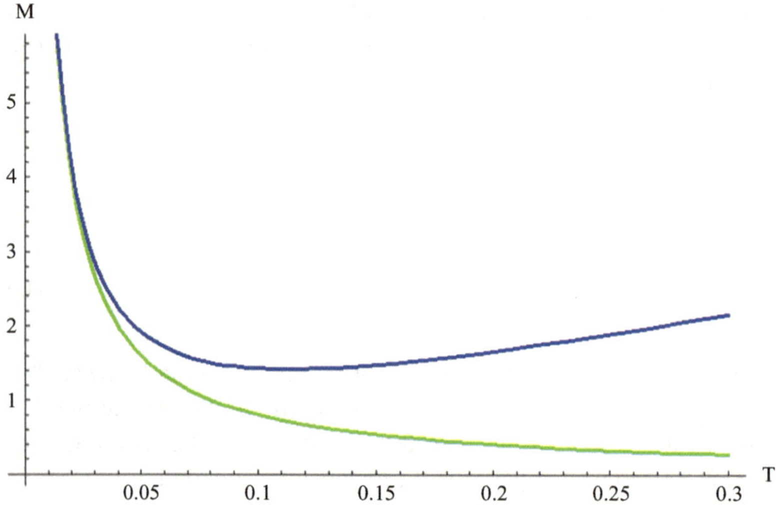

Fig. 2: Mass versus Temperature (in Planck units) for the quantum (micro) black hole.

have any thermodynamic interaction with the environment. This is confirmed by their specific heat, which drops to zero at the end of the evaporation, when $M \to M_{min}$. Remnants interact with the environment just only because they gravitate, through their final *strictly positive* mass.

5 Black Hole Remnants and Dark Matter

Exactly for the above specific property, black hole remnants have been considered, since the moment of their introduction, as very good candidates for being a major constituent of *dark matter*.

It was well known from many decades (Rubin, Thonnard, and Ford 1980) that there is an anomaly in the rotational curves of almost all the galaxies (see Fig. 3). By measuring the orbital parameters (speed, in particular) of individual stars in a given galaxy, astronomers are able to reconstruct the amount of gravitating mass necessary to generate that particular orbit. They systematically find that the gravitating mass able to reproduce the observed stellar orbits is larger (by far) than the total mass of luminous matter contained in the galaxy. So, admitting the validity of Newtonian/Einsteinian gravity, there should be present a large quantity of mass, not visible, not interacting electromagnetically, which however gravitates, and it is therefore able to explain the observed data. This invisible matter is called *dark*

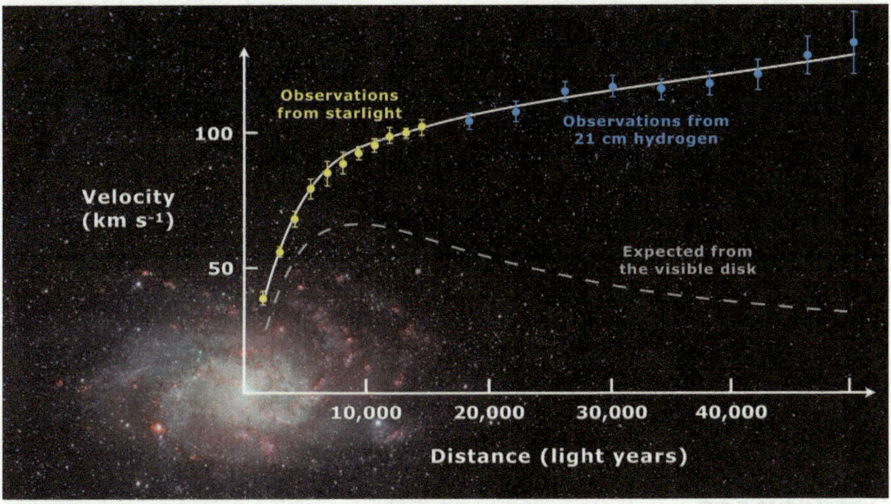

Fig. 3: Anomaly in the rotational curves of galaxies.

matter. Incidentally, the other possible explanation of the dilemma, for those who don't like the dark matter hypothesis, is to invoke a modification of the Newtonian/ Einsteinian gravity on galactic scales (Milgrom 1983; Bekenstein 2004).

During the years, many different hypotheses have been proposed as for the specific constituents of dark matter. For example, several authors pointed out that if, as it seems, neutrinos have a mass (small, but not zero), then perhaps 'clouds' of neutrinos, gravitationally bound to the galactic disks, may constitute the invisible 'dark' mass responsible for the anomalous rotational curves. Other authors propose the existence of some exotic form of matter, as axions, or maybe supersymmetric partners of ordinary particles, in order to explain dark matter.

It is therefore now clear why black hole remnants, with their specific property of having a non-zero mass, and no other interacting features, have been considered from the beginning as very good natural candidates for dark matter constituents (Adler, Chen and Santiago 2001).

6 Conclusions

This work presents a general, pedagogical introduction to the idea of how and why a fundamental principle like the Heisenberg uncertainty should be modified when the gravitational interaction is taken into account. We have therefore discussed the so-called Generalized Uncertainty Principle, and, among the many different ways, we introduce it through the idea of possible formation of micro black holes during a measurement (i.e. scattering) process. Only few of the many striking consequences of the GUP have been mentioned (also because, most of them are still a matter of debated investigations). In particular, we showed how the GUP affects a generic Hamiltonian, which in turn serves to compute modifications of several basic quantum phenomena. By comparing theoretical results with experimental numbers, upper bounds on the parameter β governing such modifications can be found. Finally, we discussed the idea of black hole remnants, one of the most distinctive predictions of the GUP applied to Hawking evaporation. Consequences of this idea for the dark matter problem are briefly mentioned. In this way, we have given a general panorama of the GUP motivations, and of some of its physical consequences.

References

Adler, Ronald J., Pisin Chen, and David I. Santiago. 2001. "The Generalized Uncertainty Principle and Black Hole Remnants". *General Relativity and Gravitation* 33 (12): 2101–2108. http://dx.doi.org/10.1023/A:1015281430411.

Adler, Ronald J., and David I. Santiago. 1999. "On Gravity and the Uncertainty Principle". *Modern Physics Letters A* 14: 1371–1381. https://doi.org/10.1142/S0217732399001462.

Amati, Daniele, Marcello Ciafaloni, and Gabriele Veneziano. 1987. "Superstring Collisions at Planckian Energies". *Physics Letters B* 197: 81–88. http://dx.doi.org/10.1016/0370-2693(87)90346-7.

Bekenstein, Jacob D. 2004. "Relativistic Gravitation Theory for the Modified Newtonian Dynamics Paradigm". *Physical Review D* 70 (8): 83509. http://dx.doi.org/10.1103/PhysRevD.70.083509.

Brau, Fabian. 1999. "Minimal Length Uncertainty Relation and Hydrogen Atom". *Journal of Physics A: Mathematical and General* 32 (44): 7691–7696. http://dx.doi.org/10.1088/0305-4470/32/44/308.

Bronstein, Matvei. 1936. "Quantentheorie schwacher Gravitationsfelder". *Phys. Zeitschr. der Sowjetunion* 9: 140–157.

Capozziello, Salvatore, Gaetano Lambiase, and Gaetano Scarpetta. 2000. "Generalized Uncertainty Principle from Quantum Geometry". *International Journal of Theoretical Physics* 39: 15–22. http://dx.doi.org/10.1023/A:1003634814685.

Das, Saurya, and Elias C. Vagenas. 2008. "Universality of Quantum Gravity Corrections". *Physical Review Letters* 101 (22): 221301. http://dx.doi.org/10.1103/PhysRevLett.101.221301.

Heisenberg, Werner. 1927. "Über den anschaulichen Inhalt der quantentheoretischen Kinematik und Mechanik". *Zeitschrift für Physik* 43 (3): 172–198. http://dx.doi.org/10.1007/BF01397280.

Heisenberg, Werner. 1930. *Physikalische Prinzipien der Quantentheorie*. Chicago: University of Chicago Press.

Kempf, Achim, Gianpiero Mangano, and Robert G. Mann. 1995. "Hilbert Space Representation of the Minimal Length Uncertainty Relation". *Physical Review D* 52: 1108–1118. http://dx.doi.org/10.1103/PhysRevD.52.1108.

Maggiore, Michele. 1993. "A Generalized Uncertainty Principle in Quantum Gravity". *Physics Letters B* 304: 65–69. http://dx.doi.org/10.1016/0370-2693(93)91401-8.

Milgrom, Mordehai. 1983. “A Modification of the Newtonian Dynamics as a Possible Alternative to the Hidden Mass Hypothesis”. *The Astrophysical Journal* 270 (2): 365–370. http://dx.doi.org/10.1086/161130.

Robertson, Howard P. 1929. “The Uncertainty Principle”. *Physical Review* 34: 163–164. https://doi.org/10.1103/PhysRev.34.163.

Rubin, Vera C., Norbert Thonnard, and Kenneth W. Ford. 1980. “Rotational Properties of 21 Sc Galaxies with a Large Range of Luminosities and Radii from NGC 4605 (R=4kpc) to UGC 2885 (R=122kpc)”. *The Astrophysical Journal* 238: 471–487. http://dx.doi.org/10.1086/158003.

Scardigli, Fabio. 1995. “Some Heuristic Semiclassical Derivations of the Planck Length, the Hawking Effect and the Unruh Effect”. *Il Nuovo Cimento B* 110: 1029–1034. http://dx.doi.org/10.1007/BF02726152.

Scardigli, Fabio. 1999. “Generalized Uncertainty Principle in Quantum Gravity from Micro Black Hole Gedanken Experiment”. *Physics Letters B* 452: 39–44. http://dx.doi.org/10.1016/S0370-2693(99)00167-7.

Scardigli, Fabio, and Roberto Casadio. 2003. “Generalized Uncertainty Principle, Extra Dimensions and Holography”. *Classical and Quantum Gravity* 20: 3915–3926. http://dx.doi.org/10.1088/0264-9381/20/18/305.

Scardigli, Fabio, and Roberto Casadio. 2015. “Gravitational Tests of the Generalized Uncertainty Principle”. *The European Physical Journal C* 75 (9): 425. http://dx.doi.org/10.1140/epjc/s10052-015-3635-y.

Scardigli, Fabio, Gaetano Lambiase, and Elias C. Vagenas. 2017. “GUP Parameter from Quantum Corrections to the Newtonian Potential”. *Physics Letters B* 767: 242–246. http://dx.doi.org/10.1016/j.physletb.2017.01.054.

Schrödinger, Erwin. 1930. “Zum Heisenbergschen Unschaerfeprinzip”. *Sitzungsberichte der Preussischen Akademie der Wissenschaften, Physikalisch-mathematische Klasse* 14: 296–303.

Gabriele Veneziano

The Big Bang's New Clothes and Eternity

Abstract: After reviewing, in simple terms, the crucial role of quantum mechanics in modern inflationary cosmology I will explain why in the inflationary scenario the traditional role of the Big Bang is completely overturned: it is played by the instant at which the Universe, *after* having been cooled down to practically zero temperature, suddenly "reheats" through an irreversible quantum process. Therefore, an unavoidable consequence of inflation is that the Hot Big Bang is neither associated with a singularity nor with the beginning of time. At the end, I will present some more speculative ideas, mostly based on string theory, on what may have preceded inflation and about whether time had a beginning.

1 Introduction and Summary

A hundred years ago two radical paradigm changes in our understanding of the physical world had already been thoroughly established:

- **Quantum Mechanics (QM)** as the correct description of the atomic world and of the quanta of light, the photons, with which atoms interact. QM marked the end of classical determinism, as emblematically expressed by Heisenberg's uncertainty principle and by Planck's constant $\hbar$ quantifying that minimal unescapable uncertainty.
- **General Relativity (GR)** as Einstein's extension of Newtonian gravity to velocities comparable to the speed of light c. Its advent marked the end not only of absolute space and time[1] but also of an absolute geometry of space-time. Its successful tests (like the deflection of light by the sun, or the precession of Mercury's perihelion) had just been carried out dissipating any initial skepticism.

For many subsequent decades both QM and GR were developed and extended, scoring success after success. But the two disciplines hardly interacted with each other. They were considered to be applicable to two very distinct physical situations: the extremely microscopic and the extremely macroscopic worlds, respectively.

As atomic experiments further developed into subatomic ones, involving electrons, protons, neutrons and other elementary particles, the necessity of combining

1 This was already the case for his previous theory, Special Relativity, which is unrelated to gravity.

https://doi.org/10.1515/9783111313610-010

QM with Special Relativity (SR) became a necessity around the end World War II. This process went on for about 30 years, culminating in the formulation, in the early seventies, of the so-called Standard Model (SM) of elementary particles and of their mutual, non-gravitational interactions. The generic name for quantum-relativistic theories, such as the SM, is Quantum Field Theory (QFT). The limitation of QFT to non-gravitational interactions was felt unimportant for a long time, given the weakness of the gravitational force among elementary particles.

In parallel, GR pursued its own adventure with striking results, like predicting the existence of gravitational waves and black holes, both beautifully confirmed recently by the direct detection of gravitational waves originating from the coalescence of two black holes to form a third one (Abbott et al. 2016). Black holes are very massive compact objects, characterized by their mass and spin, and deforming the geometry of space to such an extent that nothing, not even light, can escape from a surface (the "horizon") surrounding them.

The other class of interesting GR solutions concerned the description of the Universe as a whole, as well as its evolution, under the simplifying assumption of an approximate isotropy and homogeneity (i.e. of being roughly the same in every region and direction). It culminated in the so-called Hot Big Bang model of cosmology to be shortly described below. In all these GR developments QM was happily ignored since quantum phenomena were insignificant for the physics of the macroscopic objects of interest to GR at the time.

This parallel development of QM and GR went on undisturbed till the late seventies, early eighties. But something changed, in this respect, during the last 40 years or so: this is the topic to be discussed in this contribution.

2 Hot Big Bang Cosmology and Its Shortcomings

In order to understand why around the end of the seventies physicists working in GR and in QFT started to get closer and work together we have to recall some properties of Hot Big Bang cosmology.

Under the simplifying, but experimentally well supported, assumption of an approximately homogeneous and isotropic Universe, Einstein's equation can be solved once its matter content is given. An unexpected and startling feature of the solution is that, in general, a static Universe (one that is also the same, on average, at all times) is not allowed. The Universe must either expand or contract (or perhaps expand along some directions and contract along others). Physically, this result can be understood to be a consequence of the universal, attractive nature of gravity. An initial static Universe, left to itself, will tend to collapse. In the early twenties this looked like a serious blow for GR since people thought the Universe to be static.

Einstein was upset by this failure and ingeniously found a (we would say today "natural") solution by adding a new term to his equations, the so-called cosmological constant, Λ. Choosing properly its sign it produces a repulsive force that compensates the gravitational pull. When, in 1927, Hubble discovered, via the famous red-shift, that the Universe is actually expanding, Einstein retracted his proposal calling it his *biggest blunder* and went back to his original equations.[2]

Once we accept the expanding-Universe solution of Einstein's equations we can retrace its implications for our past history. Here we encounter another surprise: in our past the Universe was obviously smaller than today, and therefore denser. Furthermore, since Einstein's equations relate the density of matter to the curvature of space-time (meaning a deviation from our usual Euclidean geometry) the Universe was also more and more curved. The problem is that there is no mechanism, within GR, to stop and limit that growth of density, curvature (and temperature). Instead, one finds that all these quantities reach simultaneously an infinite (hence unphysical) value at a finite time in our past, the Big Bang, an event that would have occurred some 13.8 billion years ago. In jargon one calls such an event a "singularity". Finally, since the solution does not make sense before that instant, one is forced to accept that time did have a beginning, precisely at the Big Bang. This has been the standard cosmological dogma till the early eighties. And, I am afraid, it is the picture that (most) scientists are conveying to the general public even today.

Hot Big Bang cosmology had some indisputable successes. First of all it predicted, before its accidental discovery by Penzias and Wilson in 1965, the existence of a thermal bath of electromagnetic radiation, the so-called cosmic microwave background (CMB), throughout the Universe. A hot early Universe also explains how light elements, like helium and lithium, were synthesized out of hydrogen and predicts successfully their present abundance. It is therefore undeniable that the Universe, long ago, has been extremely (though not necessarily infinitely) hot. We can also locate the time of the Big Bang quite accurately: whether it was infinitely – or simply very – hot will not change appreciably the number I mentioned since all those early processes occurred in a tiny fraction of a second! As we shall see, what is not necessarily justified is to call that number the age of the Universe.

---

2 In hindsight Einstein's blunder was not so much in introducing Λ, but in having to fix it at precisely the value that leads to a static Universe. Any tiny deviation from that value would either lead to an expansion or to a contraction: in modern terminology Einstein's Λ had to be fine-tuned to an extremely high accuracy.

Actually, here is a first signal that QM can play a role in cosmology. As pointed out first by Max Planck at the beginning of the last century any quantum theory of gravity will be characterized by a length or time scale given by the appropriate combinations of c, $\hbar$ and Newton's constant G. These are called Planck length and Planck time and their values in meters and seconds are:

$$l_P = \sqrt{\frac{G\hbar}{c^3}} \sim 1.62 \times 10^{-35}\ m;\ t_P = \sqrt{\frac{G\hbar}{c^5}} \sim 5.39 \times 10^{-44}\ s;\ M_P = \sqrt{\frac{\hbar c}{G}} \sim 2.18 \times 10^{-5}\ g, \tag{1}$$

where, for later convenience, we have also introduced the Planck mass in grams.

Precisely because the Universe was so dense and curved at its birth, we can ask whether one can reliably calculate its early evolution while neglecting quantum effects i.e. by using GR as it is. Even in the absence of a satisfactory theory of quantum gravity one can argue that this is only justified from a few Planck times after the Big Bang on. In other words, statements based on GR about the Universe at times of order t_P from the Big Bang, or earlier, and at fortiori the Big Bang itself, cannot be taken seriously.

But, we may ask, what's wrong with Hot Big Bang cosmology if we decide to start our history of the Universe a few Planck times after the Big Bang and use GR from then on? This has been the pragmatic attitude of cosmologists till the seventies. Identifying what's wrong with that cosmological scenario led to the development of inflationary cosmology in the eighties.

The shortcomings of the Hot Big Bang scenario can be all ascribed to a general property of ordinary matter: the fact that gravity is an attractive force. That implies that the expansion of a Universe containing ordinary matter tends to decelerate since the gravitational attraction resists the expansion.[3] I will only mention two serious shortcoming of Hot Big Bang cosmology, both related to the above-mentioned property of the Hot Big Bang scenario.

– ***The flatness problem***

Present observations of the CMB show that today the geometry of space (on large scales) is very close to being Euclidean. However, it is easy to show that, for a *decelerating expansion*, deviation from (the flat) Euclidean geometry tends to increase with time. Inserting the appropriate numbers one finds that, unless the Universe was already Euclidean to one part in 10^{30} a few Planck times after the

---

3 Conversely, gravitational attraction would accelerate a contracting phase and this is why one gets an infinite contraction rate at the Big Bang itself.

Big Bang, it would be impossible to understand why it is now observed to be Euclidean up to, at most, one part in a hundred.

– ***The isotropy/homogeneity problem***

The CMB has a frequency spectrum typical of a thermal distribution at a temperature of about 2.7 K (meaning 2.7 degrees Celsius above the absolute zero).

As I already mentioned, predicting the existence of the CMB, including a fairly good estimate of its temperature, was one of the successes of HBB cosmology. So what's the problem? The problem is that this temperature, when measured along different directions in the sky, looks exactly the same.[4] For a *decelerating expansion,* however, one finds that the CMB radiation coming from, say, opposite directions in the sky, originated from regions of space that were *all the time* so far separated from each other that there was no way they could have ever interacted, exchanged energy, and thermalized. In other words, in order to explain the isotropy of the CMB, the newly born Universe had to be already extremely isotropic.

Both shortcomings can be best regarded as extreme fine-tuning problems, meaning that they can only be solved if one chooses initial conditions (say a few Planck times after the Big Bang) corresponding to: (i) an extremely small spatial curvature, and (ii) an extremely homogeneous/isotropic Universe.

3 A Simple Solution: Inflation!

It will not have escaped the attentive reader that the source of the above difficulties is closely related to the one confronted by Einstein when he was trying to get a static Universe from his equations: the attractive nature of gravity! In order to get an accelerated expansion one has to find some repulsive contribution like the cosmological constant Λ introduced – and then rejected – by Einstein.

Indeed, by introducing a large-enough positive Λ one does achieve, instead of a static Universe, one that has an accelerating expansion. This is, incidentally, one of the favorite explanations for the recent acceleration of the expansion as measured e.g. by the Supernovae experiments (Riess et al. 1998; Perlmutter et al. 1999). Why then not play that same game in the early Universe?

4 The first experiment to reveal that this was only true up to variations of tens of micro Kelvins (i.e. differences of about one part in a hundred thousand) was the celebrated COBE satellite experiment in 1992 (Smoot et al. 1992).

The problem is that a cosmological constant, as its name says, corresponds to an energy density that remains constant in time, in spite of the expansion. Since, instead, other forms of energy density (matter, radiation) decrease with time, an initial largish cosmological constant would have dominated every other form of energy since the early days after the Big Bang. And this is incompatible with what we know about the history of the Universe.

Fortunately however, field theories, even at the classical level, offer better alternatives. A very simple and popular one is to invoke the potential energy stored in a scalar field when the value of the field is not the one corresponding to the minimum of the potential. QFT abounds of such fields, a famous example being the Higgs field. Today the Higgs field has a non-vanishing value corresponding to where its potential energy is as small as possible. Its non-zero value generates masses for most of the elementary particles we know. But one can argue that, when the temperature of the Universe was very high, the Higgs field was actually zero. At the same time its potential energy was larger than it is today. Well, that positive potential energy behaves precisely like a cosmological constant and can produce an accelerating expansion. With an important difference: unlike a cosmological constant put in by hand, the Higgs field can evolve (e.g. as the Universe expands and cools down) and its potential energy can eventually disappear in favor of other more conventional forms of energy (e.g. radiation).

The Higgs field was just an example (yet taken as a serious possibility by some authors) to illustrate how a scalar field can play the role of a “dynamical” Λ, doing its job when it’s needed, and then kindly disappearing when it’s no longer necessary. This accelerated (quasi exponential) expansion of the Universe has been dubbed inflation, and the hypothetical scalar field responsible for it is called the inflaton.

A long enough inflationary epoch completely solves the problems we mentioned: if there were some initial deviations from Euclidean geometry (called spatial curvature in GR terminology) it is wiped out almost completely so that, even after a long decelerating expansion, it is still small (but perhaps non zero and measurable) today. By the same token, inflation can wipe out any initial inhomogeneity or anisotropy by stretching space so much that any initial ripples are now far beyond our cosmological horizon. It looks as if all problems have been cured, but that’s where QM makes its entry on the scene, as we will now discuss.

4 The Crucial Role of Quantum Mechanics

As we have just explained, inflation is very efficient for smoothing out any inhomogeneity already present at its onset. However, without the help of quantum mechanics, inflation, if it lasts long enough to solve the fine-tuning problems of hot BB cosmology, does this too efficiently: it produces an exactly homogeneous Universe.

On the other hand, in order to generate the large-scale structures we see in the sky, one has to start with a small, but finite, amount of inhomogeneities. The level of inhomogeneities we have measured in the CMB is roughly of the right order of magnitude to be able to do the job as the Universe keeps expanding and cooling down. This is again due to the attractive nature of gravity making it easy for over-dense regions to grow by accretion of surrounding matter. But what can then produce those small initial fluctuations in the CMB temperature? The mere assumption of a long inflationary phase cannot.

Quantum mechanics, instead, does just that. Although initial fluctuations are very effectively wiped out, quantum mechanics keeps creating small-scale fluctuations all the time during inflation. Like any other distance, the wavelengths of these fluctuations are stretched during the inflationary era. They are also amplified as soon as their wavelength exceeds a certain length scale (inversely) related to the energy density during inflation (which is roughly constant). This is one important free parameter: it can be called the scale of inflation l_i.

In Fig. 1 we show, in a cartoon-like style, how the initial classical perturbations (wiggles in green), as well as some quantum fluctuation (wiggles in red) produced in the earlier stages of inflation, are stretched beyond our present horizon and how, instead, quantum perturbations generated at sufficiently late times during inflation (also in red), are still within our visible Universe.

Given a specific model of inflation, the amount of fluctuations generated at different wavelength can be computed. Their magnitude is fixed by quantum mechanics in terms of the ratio of Planck's length (which, as indicated in (1), contains Planck's constant) and the above-mentioned scale of inflation l_i which, instead, has a completely classical meaning. Hence, roughly, in order to get perturbations at the level of one part in 100,000, we need a scale of inflation of about $10^5\ l_P{\sim}10^{-30}\ m.$, still a very short length scale! In turn this will fix the value of the inflaton's potential energy during inflation (which is also sometimes referred to as the scale of inflation).

Another parameter that has to be fixed is the slope of the inflaton potential: it has to be sufficiently small for inflation to last for a long time and for the scale l_i to increase very slowly during inflation (as indicated in Fig. 1). This produces an almost scale-invariant spectrum with a slight "red tilt" (a slight preference for longer wavelengths over smaller ones). In conclusion, up to adjusting a few parameters,

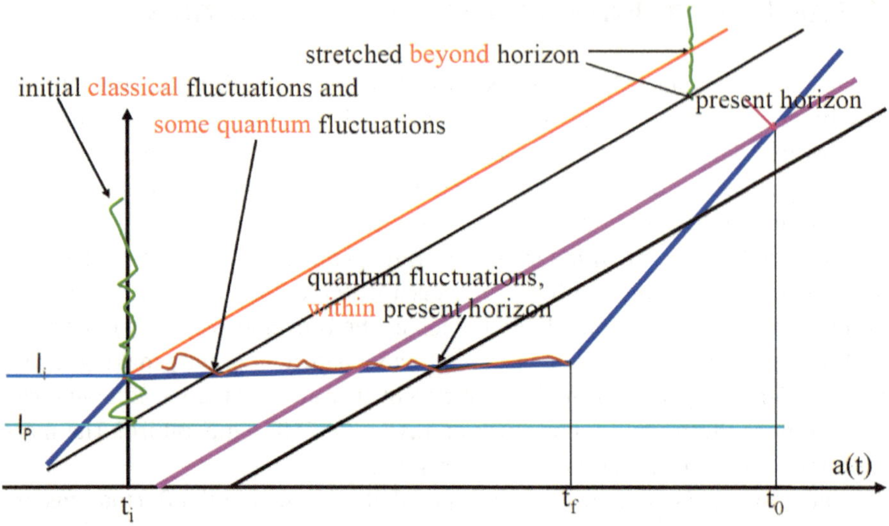

Fig. 1: Kinematics of inflation and of perturbations therein. The horizontal axis represents time in terms of the so-called scale factor $a(t)$, that tells us how physical distances are stretched by the expansion. This is why different wavelengths evolve according to straight, parallel lines. t_i, t_f, t_0 represent the beginning of the inflationary epoch, its end (to be associated with the Big Bang), and the present time, respectively. On the vertical axis we have other relevant length scales: the Planck length l_P and the "inflation scale" l_i. The thick blue line represents the horizon size (how far in space one can communicate) at different times. Its value at t_0 (our present horizon) limits the range of present observations.

inflation, together with a crucial help of quantum mechanics, can explain in an amazingly precise way the temperature fluctuations of the CMB as measured very accurately by the PLANCK satellite experiment (see Fig. 2) (Aghanim et al. 2020).

But we are not done yet . . . We need to produce the CMB itself if we want to explain its temperature fluctuations! As we have already discussed the CMB is a left-over of a Hot Big Bang. Where is the Hot Big Bang in the inflationary scenario? If it preceded inflation its consequences got wiped out by inflation. Indeed, inflation is an adiabatic expansion that cools the Universe down to essentially zero temperature. The only way out is to reheat (or just heat if it was always cold) it *at the end* of inflation so that it gets gently cooled by the later expansion.

This is the well-known reheating issue in inflationary cosmology. It is solved by dissipating the inflaton's potential energy through some quantum irreversible process, such as quantum particle creation in an external field. I like an analogy with a waterfall in which there is a lot of potential energy stored upstream of the fall, which gets converted into heat (or electricity if we are so smart to use it) as it

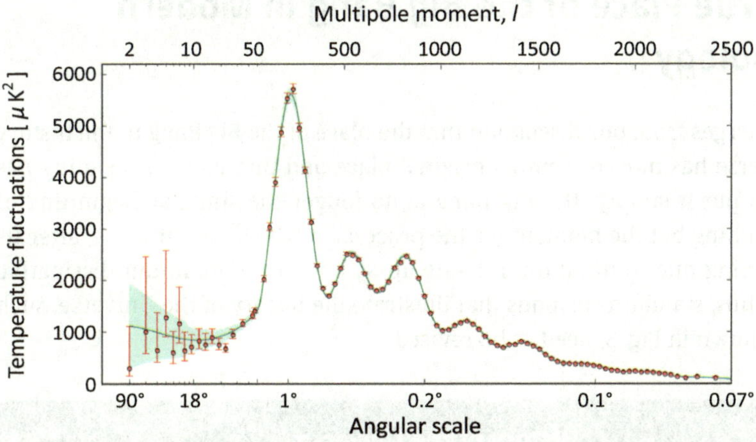

Fig. 2: The PLANCK satellite spectrum of CMB temperature fluctuations as a function of the angular between two directions in the sky (shown on the horizontal axis). With a few adjustable parameters, inflation difference accurately explains the very non-trivial structure of these fluctuations including its famous "acoustic peaks". At large angular scales statistical fluctuations prevent any reliable test.

goes down the fall. Achieving a large enough reheating temperature is one important constraint on models of inflation.

This is the second, equally important intervention of QM in inflationary cosmology. Together they produce, out of a fairly generic initial state, a Hot Big Bang with a sufficiently high (but finite!) temperature and the right amount of fluctuations to seed the large-scale structures within our observable Universe!

This line of reasoning led, around the turn of the millennium, to the formulation of what can be called the Standard Model of Gravitation and Cosmology. It is based on General Relativity, a crucial input from quantum mechanics, and the addition of two *dark sectors*. One dark sector we have already mentioned: it is the invisible (hence dark) energy responsible for the recent acceleration in the expansion of the Universe. The second, instead, is a form of massive matter that interacts with ordinary atomic matter only gravitationally. Dark matter is necessary for explaining many features of our Universe, in particular the formation of large-scale structures from the seeds we see in the CMB.

The cosmological component of the Standard Model of Gravitation and Cosmology is modestly called "The Concordance Model of Cosmology". A very appealing paradigm indeed, although we still do not know the real nature of its dark sector and we are still far from fully placing in into a grander picture of all known particles and interactions.

5 The True Place of the Big Bang in Modern Cosmology

It clearly emerges from our discussion that the place of the Big Bang in the history of the Universe has moved from its original place and that its very meaning has changed. To put it simply: the Big Bang is no longer the singular beginning of time. It is nothing but the moment (or the process) at which the Universe, after its extreme cooling due to inflation, reheats up as a result of quantum dissipative processes. Thus, standard cartoons that illustrate the history of the Universe, such as the one shown in Fig. 3, need to be revised.

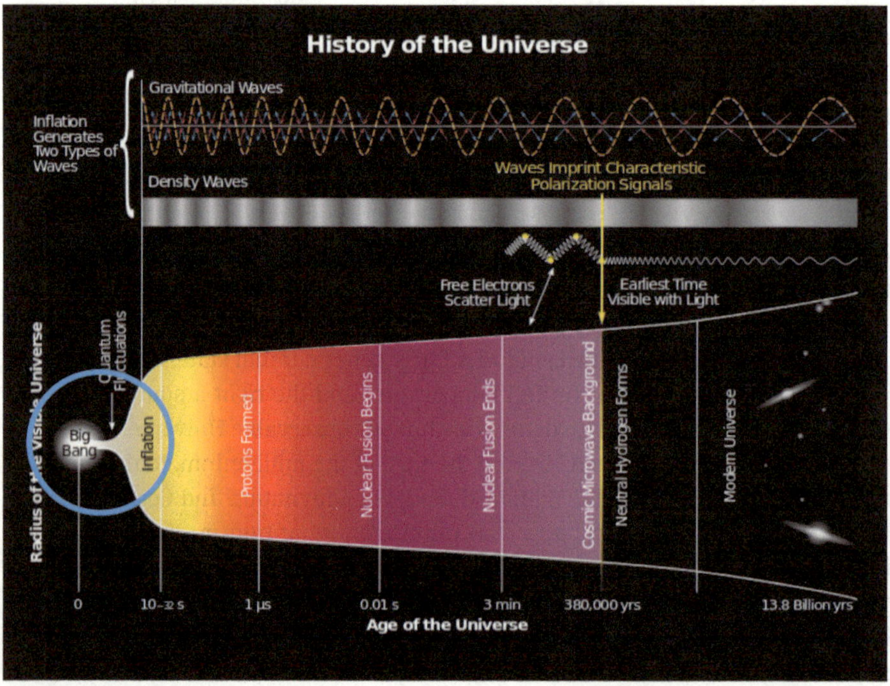

Fig. 3: A popular conventional way of depicting the history of the Universe in which the Big Bang is (mis-)placed before inflation.

Furthermore it is now legitimate, and even physically relevant, to ask: what happened before the Big Bang? And we even know part of the answer, at least for a certain lapse of time before the new Big Bang: there was an inflationary phase and we can study it through the imprint it left (by the quantum fluctuations produced during inflation) on the CMB and on the large-scale structure of the Universe. Studying what happened before the Big Bang has become a physical, no longer a metaphysical, question.

An even more direct look at what went on before the Big Bang will hopefully come in the not-too-distant future, from the detection and study of gravitational waves produced during inflation. Gravitational waves, unlike the electromagnetic ones of the CMB, traveled undisturbed even when the Universe was very hot and charged particles were not yet combined to form neutral atoms (this is why it is impossible to look at the CMB beyond the time of recombination).

The inflationary scenario (or the concordance model of modern cosmology) does not represent, however, a fully self-contained history of the Universe. In particular, they leave open the question of how, when and where inflation started.

We understand pretty well what are the conditions to be satisfied –in a certain region of space and for a certain interval of time– in order for that region to undergo a long inflationary phase and to become, today, as large as our visible Universe. We understand much less the global structure of the Universe. It could have inflationary patches of different kinds with non-inflating regions separating them. Also, each distinct inflationary patch could have different physical properties (the so-called "Multiverse").

In connection with these deeper questions too, QM is bound to play a crucial role: presumably, a full-fledged quantum theory of gravity and of the other forces will be needed in order to make progress. So far, an approximate semiclassical approximation was sufficient provided the scale of curvature during inflation was sufficiently small in Planck units. And we know that this was the case during the last phase of inflation because of the smallness of the quantum fluctuations we measure in the CMB. But, if we want to go back even further, a full quantum theory of gravity is very likely needed.

At present, the leading candidate for such a consistent theory is (super-)string theory. It is not possible, within the space at my disposal, to even try to describe in any detail what string theory is. It suffices to say that it is a candidate unified quantum theory of all forces and elementary particles based on three basic ingredients (the first two of which we have already discussed and used):
- Quantum mechanics
- Special Relativity
- The postulate that *all* elementary particles are one-dimensional objects, strings. These come in two kinds, open and closed. They are characterized by a single dimensionful parameter, the string tension T (its energy per unit length).

Note that we have not included GR in the above assumptions. In the same way that one can define a Planck length, a Planck time, and a Planck mass via equation (1), one can define a string length, a string time, and a string mass by:

$$l_s = \sqrt{\frac{c\hbar}{T}}; \quad t_s = \sqrt{\frac{\hbar}{cT}}; \quad M_s = \sqrt{\frac{\hbar T}{c^3}}. \tag{2}$$

Notice that these "string quantities" go to the corresponding Planckian ones if we identify the string tension T with c^4/G. Is this the way string theory represents gravity? Not quite. In string theory the quantities appearing in (2) are the most basic ones, as we shall see in a moment.

By contrast, Newton's constant G (and thus Planck's time, length and mass) are not as fundamental. They are related to the string quantities through the so-called string coupling g_s, a dimensionless number that controls the strength of all interactions. It is roughly related to the famous fine structure constant $\alpha \sim 1/137$ of QED. Because of this value of g_s the string length is not expected to be more than a couple of orders larger than the Planck length[5] but this is sufficient for it to have a huge impact on quantum gravity. At the string mass/energy scale all four forces unify and have a strength given in terms of g_s^2. At low energy string theory is well described by its "massless modes" hopefully to be identifiable with the particles of the standard model or of some extension of it, while at high energy and short distances it will differ from an ordinary quantum field theory in essential ways.

The existence of these massless modes is of course phenomenologically crucial. It is actually a consequence of QM since, classically, the only massless string would be point-like. Instead quantum mechanically a string can have a finite size and yet be massless and even carry non-zero angular momentum. These latter strings of spin up to 2 would represent the carriers of all known long-range interactions (including gravity mediated by the spin-2 graviton, this is why GR comes out of string theory!) and possibly more . . .

The finite size of strings, on the other hand, manifests itself through modifications of the predictions of QFT when scales of order l_s or shorter are probed. This is precisely what happens when the Universe, as described by GR, has an age of the order of t_s. Since $t_s > t_P$ this means that string modifications intervene *before* one enters a Planckian regime.

---

5 For the learned reader: there is here an amusing analogy with the theory of weak interactions where there is a fundamental mass scale given by the $W-Z$ masses and another, more phenomenological one, associated with Fermi's constant G_F. Also there the ratio of the two scales is given by a coupling whose value is not very far from 1.

Several *gedanken experiments* have been considered in order to study this "stringy" regime and have confirmed the softening of GR expectations at short distances. They are sometimes described by an effective Generalized Uncertainly Principle (GUP) reading (Amati, Ciafaloni, and Venziano 1989)

$$\Delta x \geq \frac{\hbar}{\Delta p} + \frac{c\Delta p}{T} \geq 2l_s, \tag{3}$$

in which the first term is the usual Heisenberg's uncertainty, the second its "stringy" extension, and the last inequality is easily proven. Here we see l_s emerging as a minimal measurable length. Armed with these rough notions about strings, let's go back to the question we asked earlier.

6 Before Inflation . . .

General theorems proven by S. Hawking and R. Penrose in the seventies state that, within General Relativity, an initial singularity is unavoidable. However, "before" (as we go backward in time) reaching the singularity we necessarily encounter a situation in which certain physical quantities (density, temperature, curvature of space-time) reach values of $O(1)$ when measured in the string units of (2). Because of the (small but sufficient) hierarchy between string and Planck units, string effects intervene *before* one reaches the regime in which quantum gravity corrections go out of control. String effects, on the other hand, are well known to tame the bad behavior of quantum gravity at short distances (solving e.g. its "non-renormalizability" problem).

It is also known, for instance, that string theory sets an upper limit (of course of O(1) in string units) to temperature (the so-called Hagedorn temperature). And even measuring distances (and time intervals) smaller than l_s (or t_s) don't make sense in view of (3).

We may thus reasonably guess that, within string theory, there is no place for a singularity taking place before inflation. Rather, the singularity and what follows it for a time $O(t_s)$ should be replaced by a *stringy phase* during which a conventional classical description of space-time is no longer valid. String theory allows for solutions that do not correspond to any smooth classical geometry and yet are perfectly well defined. One such solution could represent the true beginning and evolve at later times into a more conventional inflationary epoch.

There is, however, an interesting alternative to this possibility within string theory, going under the name of Pre Big Bang (PBB) scenario. Born in the early nineties it has been the prototype of a whole class of cosmologies now known as *bouncing* cosmologies. In this class of scenarios, the usual cosmological expansion (with or without an inflationary epoch) is preceded by a contracting phase which, through a bounce, turns into an expansion. Such a behavior is forbidden in Einstein's gravity but quantum/string effects could possibly induce such a bounce.

Many scenarios of this sort have been proposed, but I'll limit myself to describe the original one since it is deeply rooted in some novel symmetries characteristic of string theory. These symmetries, known as *T*-dualities, involve in an essential way a scalar field which is ubiquitous in string theory as an inevitable partner of the gravitational field: it is called the dilaton and plays a very important role in string theory. The above-mentioned string coupling g_s is not a God-given number but is itself a field, the dilaton. Even if we know, through its connection with the fine structure constant, that the value of this fields cannot have changed appreciably since many billions years, it is all but excluded that it may have evolved in primordial cosmology and, a fortiori, before the Hot Big Bang.

Actually the symmetries of string cosmology associate with the usual decelerated expanding solution one of accelerated expansion (i.e. of inflation), the acceleration being driven by the evolution of the dilaton from the regime of extremely weak coupling to the one corresponding to its present value. This pre-bounce phase has been dubbed dilaton-driven inflation (DDI). While the coupling grows, so does the ratio l_P/l_s, so that, while the size of the Universe keeps growing all the time in units of l_s, it undergoes a contraction before the bounce if sizes are measured in units of l_P. What remains true, independently of the units adopted, is that the curvature of space-time grows before the bounce and decreases afterwards: the curvature undergoes a bounce! Details of the bounce itself will only be describable in a full quantum string theory context to be still fully developed.

A very ambitious possibility is that the DDI phase plays the role of ordinary inflation so that, after the bounce, one goes over directly to a standard Hot Big Bang cosmology (explaining the name given to this scenario). Quantum particle production during the pre-bounce phase has been shown to be able to heat-up the initially cold Universe. A detailed scenario has been constructed (see Gasperini and Veneziano 2003 and references therein), invoking other fields present in string theory, which is compatible with CMB observations and predicts a vanishingly small *B*-mode in the CMB polarization. It may also be able to generate seeds for the observed cosmic (intergalactic) magnetic fields whose origin is still very mysterious.

On the negative side, the PBB scenario can (but does not automatically) generate a quasi-scale-invariant spectrum of perturbations and may also have difficulties in washing away certain kinds of initial anisotropy. Therefore, a more modest possibility would be that the pre-bounce phase is followed by the above-mentioned string phase and, finally, by a long enough conventional inflationary epoch.

Of course all of this is highly speculative. The good side of the story is that it is no longer a tabu to ask experimentally questions like: What happened before the Big Bang? What happened before inflation? One day we may even find out, scientifically, the answer to a philosophical question that goes back to Augustin, Aristotle, and probably much farther in the history of mankind, the question of *eternity* mentioned in my title: *Did time have a beginning?*

References

Abbott, Benjamin P., Richard Abbott, Thomas D. Abbott, Matthew R. Abernathy, Fausto Acernese, Kendall Ackley, Carl Adams et al. 2016. "GW151226: Observation of Gravitational Waves from a 22-Solar-Mass Binary Black Hole Coalescence". *Physical Review Letter* 116(24): 241103. https://doi.org/10.1103/PhysRevLett.116.241103.

Aghanim, Nabila, Yashar Akrami, Mark A.J. Ashdown, Jonathan Aumont, Carlo Baccigalupi, Mario Ballardini, Anthony J. Banday et al. 2020. "Planck 2018 Results. VI. Cosmological Parameters". *Astronomy and Astrophys* 641: A6 1–67. https://doi.org/10.1051/0004-6361/201833910.

Amati, Daniele, Marcello Ciafaloni, and Gabriele Veneziano. 1989. "Can Space-Time Be Probed Below the String Size?" *Physics Letters B* 216: 41–47. https://doi.org/10.1016/0370-2693(89)91366-X.

Gasperini, Maurizio, and Gabriele Veneziano. 2003. "The Pre-Big Bang Scenario in String Cosmology". *Physics Report* 373: 1–212. https://doi.org/10.1016/S0370-1573(02)00389-7.

Perlmutter, Saul, Gregory Aldering, Gerson Goldhaber, Richard A. Knop, Patricia G. Castro, Susana Deustua, Sebastien Fabbro et al. 1999. "Measurements of Ω and Λ from 42 High Redshift Supernovae". *The Astrophysical Journal* 517(2): 565–586. http://dx.doi.org/10.1086/307221.

Riess, Adam G., Alexei V. Filippenko, Peter Challis, Alejandro Clocchiatti, Alan Diercks, Peter M. Garnavich, Ron L. Gilliland et al. 1998. "Observational Evidence from Supernovae for an Accelerating Universe and a Cosmological Constant". *The Astronomical Journal* 116(3): 1009–1038. http://dx.doi.org/10.1086/300499.

Smoot, George F., Charles L. Bennett, Alan J. Kogut, Edward L. Wright, Jon Aymon, Nancy W. Boggess, Edward S. Cheng et al. 1992. "Structure in the COBE Differential Microwave Radiometer First Year Maps". *The Astrophysical Journal* 396(1): L1–L5. http://dx.doi.org/10.1086/186504.

Giacomo Mauro D'Ariano

For a Science of Consciousness

Abstract: The ultimate motivation for our greed for knowledge on how the world is made, is also the unacknowledged desire for a solution to the most compelling and inexplicable mystery that grips us: death! What will become of my "self" after death? Religion provides an answer, which, however, is not shared. Instead, we would like science to provide an objective answer. So why not take the bull by the horns and face directly, on scientific grounds, what it means to "feel alive". What is "awareness"? Is it possible to scientifically deal with the problem of "consciousness"? We do not require to solve the problem, but, at least, to know if we can say something about it on the basis of shareable assumptions, based on our knowledge of the physical world, and hopefully, with some experimental confirmation.

1 Introduction

So, let's start with "how the world is made".

There are two fundamental theories that rule the world: quantum theory and classical theory. By these two names I mean the theories that provide the "grammar" for the corresponding "mechanics" of particles and fields, that is, the theories that, rather than explaining how the world is made, provide us with the rules for how to explore the world. Both theories, classical and quantum, are in all respects "information theories". What information are we talking about? We are talking about classical information, the one that we process inside our computer, and that we communicate by our cell phones, on the Internet. But we are also talking about a new type of information – the "quantum information" – which we process in "quantum computers" – a new type of computer that you have surely heard about. This is the same quantum information that we use in quantum cryptographic communications, which ensure security based on physical laws rather than on algorithmic complexity.

Some of you will ask, "What does information have to do with physics?" The great Richard Feynman and his mentor John Archibald Wheeler strongly advocated the information paradigm of physics: "physics as information". And they were right! We demonstrated this in Pavia with my research group. In short, I can tell you that the quantum theory of non-interacting fields – namely the fields of Weyl (neutrino), Dirac (particles with mass) and Maxwell (photons) – are derived from very simple algorithms called "quantum cellular automata". And,

https://doi.org/10.1515/9783111313610-011

most surprisingly, the specific automata that give rise to the theories of Weyl, Dirac, and Maxwell are the simplest nontrivial ones that can be considered! I won't go into the technical details. To make things understandable, one can have an intuition by imagining "space" as an infinite screen with three-dimensional pixels, in which each pixel is a quantum bit (the so-called "qubit"), and physics emerges from that screen. The pixel is extremely small, of the order of the Planck length: 10^{-33} cm corresponding to a size so tiny that if we enlarged it up to 1 millimeter, we would see one centimeter enlarged to one million light years, equal to the diameter of an entire galaxy!

2 Quantum and Classical Information

Indeed, the actual situation is more complex. In short, what's happening "out there" – in what we call the "world" – is both quantum and classical information processing, with quantum information transforming into classical and, conversely, classical information transforming into quantum. An example of transformation is what we have in the lab, where we transform quantum information into classical when we perform measurements, and, vice versa, we transform classical information into quantum when we prepare particles or anything else on which we perform a quantum measurement. In summary: we transform classical information into quantum one when we perform "preparations", and transform quantum information into classical one when we perform "observations".

The main difference between classical and quantum information is that we can "see" classical information, while we cannot see the quantum one. For example, in a Stern-Gerlach experiment we see the source emitting particles and we see the final traces of the particles on the target screen, but we cannot see them flying in between. I emphasize that "we only see classical information", we cannot see quantum information! Quantum theory provides what "theoretically happens" between what we see at the input and what we see at the output, but we cannot see quantum objects "flying" in between. Quantum theory connects the classical events that we see, through quantum entities that we cannot see, and which are simply theoretical entities. This fact opens up the "problem of realism" of quantum theory: the entities we theorize cannot be seen! Quantum theory introduces invisible connections between visible objective events!

But let's ask ourselves: is there really a logical need to "see" all theoretical objects? Not really! However, such a limitation leaves some physicists unsatisfied. (By the way, we do not actually see many notions of classical mechanics, such as energy, gravitational and electromagnetic fields, etc.)

3 The Problem of Consciousness

But let's go back to our problem of interest: how to face, as a physicist, the problem of "consciousness" or "awareness". Let's see what this has to do with what I have said so far.

In the theory that I proposed with Federico Faggin (D'Ariano and Faggin 2022), consciousness is quantum (and I will immediately explain why). Therefore, the outer world that we see is classical, while the inner experience inside us is quantum! Outside of us we see only classical objects, while internally we live experiences and thoughts, which are quantum! Note that nothing is more real to us than our direct experience: some neurologists assert that consciousness is an "epiphenomenon". This is equivalent to declare that the only thing we are sure of, our personal experience, the Descartes's "cogito ergo sum", does not exists. On the contrary, the external world (which is classical) in principle could only be an illusion: we could be a "brain in the vat", like that of Hilary Putnam, namely, a brain immersed in a tank full of physiological liquid, connected through many wires to a computer that is simulating a virtual reality. The only thing I know for sure is that "what's out there" is distinct from me: it is not me, simply because what is "out there" happens independently of my will. We therefore have an inverted situation between the "inside" and the "outside". "Inside": what we live and feel is quantum. "Outside": what we see is classical, whereas what is quantum cannot be seen!

Why should consciousness be quantum? There are many reasons for such hypothesis.

Reason one: if we assume that the "intimacy" of consciousness is an "in principle" feature (i.e. the intimacy of our thoughts would resist any future technology), well, . . . this is exactly the case of quantum information! In quantum theory the principle of "no information without disturbance" holds. The principle states that: if I try to read quantum information I inevitably modify it, and in such a way that it is impossible to determine the quantum state using subsequent readings. On the other hand, a theorem due to Holevo states that no measurement (even made up of many different measurements performed on the same system) can provide more information than a single classical bit per quantum qubit. Recall that, unlike the bit that has only two states – "zero and one", the quantum bit – or "qubit" – has a continuum of states corresponding to all points of a sphere (which we call the Bloch sphere). Thus, from a continuum of possible states we cannot obtain more than a single bit of classical information! And, to complicate matters, the measurement irremediably changes the state of the qubit, thus denying the possibility of obtaining more information through additional measurements carried out in sequence on the same system. We conclude that anyone who attempts to share our internal experience can get only very little classical information, and at the cost of completely destroying our own conscious experience!

4 Information Sharing

The shareability of quantum information is impossible, whatever is the method one devises, for example by making many copies of the same qubit, since this would violate the no-cloning theorem. Quantum information cannot be copied (by the way, this is the basis of its information technology security!) It is useless to try to reduce the disturbance produced by the copying of quantum information: we would in any case violate the quantum trade-off between information and disturbance, namely: the less we disturb, the less information we get; the more information we get, the more we disturb! We conclude that our quantum experience cannot be shared in any way, not even in principle. Only classical information can be shared. Quantum information is transformed into classical (and, as already mentioned, no more than one bit per qubit) when we store the experience: the information is quantum only in the short-lived "buffer" necessary to build our overall experience. It is crucial that long-term memory be classical: indeed, this solves the criticisms that are often made based on the fact that interactions in the brain in the long run would destroy quantum coherence (which, indeed, was the critique made of the quantum consciousness proposal of Penrose). Therefore, long-term memory must be classical to survive random interactions in the brain! Now, since at most one bit of classical information can be extracted from a qubit, the memorization process (and subsequent "recall" from memory) considerably degrades the experience stored in the transformation from quantum to classical and back to quantum. The optimal capacity of the store-and-recall channel was evaluated by myself theoretically, and in principle it can be confronted with cognitive experiments.

5 Inner Experience Is Quantum

But let's ask again: "why is it plausible that the inner experience is quantum?" We have seen that one answer is the consideration of its in-principle intimacy. The second answer is the possibility of constructing "qualia" and therefore "thoughts" through quantum superposition and entanglement of qualia.

For those who do not know what "qualia" (plural of "quale") are, in philosophy of mind these are the qualitative aspects of conscious experiences. They are ineffable because they are related only to the subject who experiences them, who cannot tell others how he is seeing, tasting, smelling, etc. For example, consider the "quale" of the color: it will never be possible to check if my color scale is inverted from purple to red compared to that of another person who has the colors in the opposite order, namely he sees violet in place of red.

But let's go back to the second motivation of the quantum nature of experience: the possibility of building "qualia" and "thoughts" through quantum superpositions and "entanglement". Let's consider the case of a spin which is also an information qubit. According to quantum theory, by superimposing the spin-up (directed upwards) with the spin-down (directed downwards), if we make the sum of the two we obtain a "spin-left" (namely directed towards the left), whereas if we make the difference we obtain a "spin-right". That is, by superpositions we generate new direction-qualia. This looks like what happens when we overlay colors, for example green and red to create yellow etc. With entanglement we can overlap two or more qualia creating completely new qualia, in exponentially large number, and with only a few qubits. And by composing qualia with qualia we build up thoughts. Everything in our inner experience is made of qualia. Sometimes qualia seem to be located in our mind, sometimes inside our body (like pain or pleasure), or they do not have a specific place (like happiness or sadness). We have possible quantum states of qubits that function as a variable color-sound-odor-etc. qualia single-LED on the "screen" of our consciousness, while classical bits would be black and white. We can therefore consider consciousness as "the ultimate screen", operating with a special type of information: quantum information. Information processing involves our decisions based on input, namely the so-called "free will". But let's go into more detail in the theorization of consciousness.

6 A Theory of Consciousness

In proposing our theory, Federico Faggin and I followed the line suggested by David Chalmers to solve what he calls the "hard problem", because it cannot be solved in terms of a "functioning", e.g. of a part of the brain, but requires explaining "the experience associated with the functioning". To clarify this difference, Chalmers introduced the "zombie" paradigm, that is a being who, from the outside is perfectly indistinguishable from a human being, but, from the inside, has no experience, doesn't feel anything! Chalmers wonders "why is there that feeling that accompanies awareness of sensory information?" The difference between the hard problem and the easy problem is that the "cognitive" easy problem can in principle be solved through the usual "physicalist" scientific strategy, while this is not possible for the hard problem. Chalmers argues that there is an "explanatory gap" from the objective to the subjective.

Following Chalmers' suggestion (Chalmers 2010), Federico Faggin and I started from a solution that, leaving physics untouched, proposes an irreducible psycho-informational postulate to solve the hard problem, that is the hypothesis that a fundamental property of "information" is its experience by the physical "system" that supports the information itself. In short, the qubit "feels" its quantum state. This requirement does not interfere at all with physics, it does not require modifications of it, but only adds a further property of "information", namely that information is "experienced", "felt", "lived" by the system that supports it.

So, as I mentioned, the type of information involved in consciousness must be quantum for multiple reasons, including its inherent privacy and its power to construct complex thoughts by entangling qualia. The resulting theory is a panpsychism based on quantum information from which consciousness follows.

We therefore argue that the quantum state experienced internally, since it corresponds to a definite experience, must be a "pure" quantum state, which we call "ontic", and not the probabilistic catalog of possible states that an external observer might infer, a state that we call "epistemic", which is generally a probabilistic mixture of pure states possible "a priori".

The purity of the ontic state requires an evolution that preserves its purity, that is to say a so-called pure "quantum operation". This transformation, being quantum, occurs non-deterministically, and its particular outcome is interpreted as "free will", in how unpredictable it is, even in principle, since quantum randomness cannot be interpreted as a lack of knowledge of something pre-existing, but represents a real "act of creation".

As we have seen, in this way we arrive at a form of panpsychism. However, the very purity of the state, as well as its evolution, allows us to solve the well-known "combination problem" – the main problem of panpsychism. In summary, the combination problem is the problem of whether and how several conscious minds can be composed to constitute or give rise to a new conscious mind. This problem is solved by entanglement. In fact, if I consider two individuals in a pure joint state, since such state is generally entangled, quantum theory predicts that each of the two individuals would result in a "mixed" state – not pure as required for the ontic state. Furthermore, individuality is connected with experiential continuity, and with memory, which, as already said, is classical. The evolution of the quantum state represents only a short-term experiential buffer, and itself contains information transfers from quantum to classical, and from classical to quantum. Long-term memory, on the other hand, due to inevitable decoherence, cannot be quantum. Therefore, since memory is classical, conscious experience needs memorization and subsequent rereading processes to go from quantum to classical and from classical to quantum, with a substantial loss of information. Such memorization and recall processes can benefit from having multiple copies of the experienced state re-prepared via what we call "attention", and thus allowing for a better quality of classical memorization. In principle, such information loss can be subjected to experimental verification.

7 Operational Probabilistic Theories and Awareness

More general information theories than quantum and classical theories could also be explored: these are what we now call OPTs, or "operational probabilistic theories", a large class of information theories that contain classical theories and quantum. However, it appears that the theories of quantum and classical information together give rise to a logically compliant and self-sufficient theory of consciousness. The advantage of operational theory (like classical and quantum theory) is that it represents a "black-box" theoretical approach, independent of physical or physiological implementations. And, in any case, a theory of consciousness must be an information theory, namely an OPT.

Information is everywhere, not just on the Internet and on media. Light hits objects and subsequently reaches our eyes, providing us with information about those objects: color, position and shape. The information is supported by a series of different systems: the modes of vibration of light, followed by the retina, then by the nerves and finally by the neurons. Consciousness – namely the David Chalmers' "hard problem" – is resolved in terms of the direct "experience" by the final system that supports information. It is the most direct use of a very structured kind of information, which manifests itself through different types of qualia: colors, sounds, flavors, smells, touch, sensations, pain, pleasure, sadness, happiness. We can say with great confidence that awareness is the ultimate manifestation of a type of information: quantum information. As a conscious entity, each of us is a system that supports this information. This is, in summary, the starting hypothesis, namely:

Awareness is "being the system" that supports information. It is "the feeling of information processing". "Being the system" allows you to "feel" the information and its processing directly.

8 Conclusions

And, in ending my paper, I would like to return to the ultimate motivation of our greed for knowledge of how the world is made, that is, the desire for a solution to the mystery of death. Albert Einstein in a letter to the sister of his dear friend Michele Besso on the occasion of his untimely death wrote:

> Michele has left this strange world a little before me. This means nothing. People like us, who believe in physics, know that the distinction made between past, present and future is nothing more than a persistent, stubborn illusion (Einstein 1955).

This is the so-called "Block Universe" worldview, namely: the passage of time is only an illusion, and the "present" is simply the equivalent of a temporal positioning in space-time. For Einstein, Past, Present and Future are completely determined, since they are deterministically connected, on the basis of reversible physical laws, that is, on a reversible and eternal space-time, where temporal incursions are possible, because everything is determined.

Unfortunately, according to quantum theory – the theory that Einstein contributed to, but never "digested" – this is not the case! In fact, the correct interpretation of quantum randomness is that the result of the experiment is, literally, an "act of creation", not the reading of something pre-existing. Quantum theory is a theory of irreversible processes, because the quantum experiment cannot be reversed, from the result to the preparation of the physical system on which the experiment is performed. More precisely, the experiment can be reversed with a certain probability, but at the price of canceling the information given by the previous experiment, because the Shannon's information attained from the second measurement contradicts the information achieved from the previous measurement. In quantum theory there is no information that derives from reversible transformations.

We deduce that death, alas, is not reversible (except with a ridiculously small probability), contrarily to what Einstein tried to make Michele Besso's sister believe, namely that it is possible to roam around at will in time as well as in space. We conclude that, as a person, each of us is an "unicum" in the history of the world. Are we reborn? This cannot be excluded: but, if it were the case, it would certainly be in different guises.

Quantum theory introduces the "act of creation" into physics, the "authentic change". Parmenides said that "reality never changes", as in the "Block Universe", where everything is already written in space-time. Heraclitus said that everything changes constantly, and that "there is no permanent reality except the reality of change: permanence is an illusion of the senses". Quantum theory proves Heraclitus right. Heraclitus also said that "nature loves to hide". He was right in this too: quantum loves to hide!

References

Chalmers, David J. 2010. *The Character of Consciousness*. Oxford: Oxford University Press.

D'Ariano, Giacomo Mauro, and Federico Faggin. 2022. "Hard Problem and Free Will: An Information-Theoretical Approach". In *Artificial Intelligence versus Natural Intelligence*, edited by F. Scardigli 145–192. Cham: Springer. https://doi.org/10.1007/978-3-030-85480-5.

Einstein, Albert 1955. *Letter from Albert Einstein to his friend Michele Besso*. In World Collectors Net. https://www.worldcollectorsnet.com/auction-news/letters-albert-einstein-friend-michele-besso/, last accessed 23 October 2023.

Federico Faggin

Freedom and Artificial Intelligence

Abstract: Artificial intelligence (AI), now promoted as a portrait of our nature and a forecast of our future, is mirroring a reality devoid of any human values. This is bringing into the foreground questions that science, philosophy, and religions have never properly answered: Does free will and consciousness exist? And if so, how can their existence be reconciled with physics that can so successfully predict and explain the functioning of the physical world? But how can physics explain the concepts of free will and consciousness when they barely exist in its vocabulary? This article will describe a rationale and a theory that can finally give sensible answers to these crucial and existential questions.

1 Introduction

The recent successes in artificial intelligence (AI) have caused an upsurge of "digital ontology", a physicalism based on information. We are repeatedly told that we are biological information processing systems no different in essence than a robot made of metal, plastic, and silicon. The word *wetware* is often used in a derogatory tone since silicon hardware can do billions of multiplications per second, compared with just one in a few hundred seconds, the best we can do. Many engineers are saying that soon computers will become conscious and will be smarter than us. Neuroscientists insist that when we think we made a free-will decision, our brain-computer has already decided for us and is simply informing our consciousness of its decision.

Physicists and biologists are happy to hear these news since they have been saying for the longest time that we are classical machines like computers and therefore free will is an illusion. Everything seems to fit perfectly with the materialistic credo, and many philosophers and sociologists maintain that it is time to let go of religions and spirituality and embrace the brave new world we have in front of us. Everything is information and the future is bright!

What is true in this picture? Does free will exist? Are we really machines? What is the role of consciousness? Is there any purpose to the universe and to life? Is this digital ontology sound?

I am a physicist, inventor, and entrepreneur who has taken a serious look at the fundamental nature of reality after a spontaneous extraordinary experience of consciousness more than 30 years ago that gave me a glimpse to a vaster perspective than what we develop in our ordinary experience. I experienced myself

https://doi.org/10.1515/9783111313610-012

as the universe observing itself with my own point of view! (Faggin 2021, 157–161) What made this event astonishing was its impossible perspective because I was both the experiencer and the experience. I was simultaneously the observer of the world and the world; I was the world observing itself! And I was concurrently *knowing* that the world is made of a "substance" that feels like love and that *I am* that substance!

This was a form of *direct knowing* even surer than the "certainty" offered by logic. It was a knowing from the inside rather than from the outside that simultaneously involved all aspects of my consciousness: the physical, emotional, mental, and spiritual levels. For the first time in my life I experienced the existence of another dimension of reality, the spiritual level, in which a person feels one with the world. This dimension cannot be known intellectually by reading a book. It can only be known by living it, *by being it*.

This awakening to an unknown and powerful nature clearly showed me that there was much more to consciousness and life than the ordinary, mechanical reality described by materialists, despite the many accounts similar to mine that are simply disdainfully dismissed, if not ridiculed.

I decided to find out for myself, beginning with an intense investigation that brought me many other extraordinary and spontaneous experiences of consciousness that revealed many other unsuspected properties tucked away in our deepest nature. Most importantly, I realized the immense power of suggestion that can trap us inside a conceptual prison with walls built and solidly held in place by beliefs that are never tested. Materialism is such a prison because it discourages straying outside its credo, insisting that we are machines without meaning and purpose, encouraging the avoidance of emotions, the distrust of intuitions, and the belief that only objective, measurable reality is real.

I then began to question the theory that describes us as biological machines similar to computers because, simply based on the known laws of physics, we should be completely unconscious, just like our computers are. In fact, the scent of a rose and the love for a child should not exist because no scientist can explain how electrical or biochemical signals can produce these qualia. We know only because we have a conscious experience, and this fact seems to be irrelevant to those who believe that inanimate matter can explain all of reality. To them, when something cannot be explained within their dogmatic theory, it can be declared epiphenomenal and therefore there is nothing to worry about. This is exactly what has happened to consciousness and free will.

As I progressed in my study, I gradually realized that consciousness and free will may be irreducible properties of nature since I could explain how unconsciousness could emerge as a special case of consciousness, and how determinism could emerge as a special case of free will, but not vice versa, for how can "more" emerge from "less"?

The end result of this 30-plus years of research is a theory of consciousness and free will based on the nature of quantum information that G.M. D'Ariano and I recently developed (D'Ariano and Faggin 2022, 145–191) and I will describe later in its essential lines. This theory radically changes our worldview because, in it, consciousness and free will have always been part of One, the totality of what exists, and thus the evolution of the physical universe must have been dramatically influenced by their presence. Within this theory we are emanations of One, and each of us contains the essence of One. The physical universe is only the symbolic aspect of a vaster semantic-symbolic reality and AI is just a mechanical configuration we have imposed upon classical matter. In this theory, AI stands in the same relationship with human intelligence as classical information (bits in space-time) stands with quantum information (entangled qubits in Hilbert space).

2 Consciousness and Free Will

Our consciousness is the inner semantic space in which the signals coming from the physical world and processed by the brain take the form of feelings, sensations, and meaning, i.e., what philosophers call *qualia*. Qualia refer to what it "feels like" to have a conscious experience. Note that the nature of feelings is completely different from the nature of physical events. Physical events happen in space-time and are accessible "from the outside" through our physical senses and instruments, and produce a so-called third-person experience common to all observers. A feeling is instead a *private,* first-person experience accessible only "from the inside" by the owner of the consciousness. Note that the interiority I am speaking about, is not a physical dimension.

Let's consider, for example, how a rose is recognized by its scent (Faggin 2021, 154). A rose emits particular molecules with unique three-dimensional structures. They can enter as "keys" into the "locks" of some receptor molecules incorporated in the membranes of the olfactory cell of the nasal epithelium. When this happens, the olfactory cells produce macroscopic electrical signals that constitute the input to the neural networks of the olfactory cortex. The output of these networks are signals that correspond to the name of the identified object: *rose.*

Even a machine can recognize a rose by its molecular "emissions", emulating the natural process I just described. However, this recognition is *qualitatively* very different from ours: a machine feels nothing, and the name of the recognized object is just another blind signal or symbol. The artificial nose does not have the conscious experience of the fragrance of the rose, and the way in which it will respond to the “rose” symbol depends solely on how it is programmed. For us, however, the scent of the rose is a *quale*, something completely different from the electrical signals generated by the neural networks. It is correlated to them, of course, but it is not identical, nor can it be produced directly by them since it has a completely different quality from the electrical or mechanical activity that classical signals entail.

The scent of a rose, just like the taste of cherry jam, the sound of a violin or the feeling of love, is a *feeling* that makes the subject consciously aware of the symbolic data. It constitutes an *experience* taking place in our consciousness that can deeply affect us and brings with it the *comprehension* of the situation. The computer, on the other hand, cannot be aware of anything. Therefore, the comprehension brought by consciousness is not available to a computer. And herein lies the fundamental limit, and the danger, of artificial intelligence.

But why are we not processing information in the dark like a computer? What benefits do we get from consciousness? What produces qualia? And how does the conversion from signals to qualia occur? Science cannot answer these questions.

For years I unsuccessfully labored to understand how consciousness could arise from electrical or biochemical signals, and I found that, invariably, electrical signals can only produce other electrical signals or other physical consequences such as force or movement, but never sensations and feelings, which are qualitatively different.

I therefore came to the conclusion that consciousness has to be a fundamental property like electricity, for example, that could not possibly arise from elementary particles devoid of electrical charge and magnetic spin. In other words, consciousness must be an irreducible property of the “elementary particles” of which everything is made, just like the electrical charge is. Consciousness must be a fundamental property that does not derive from any simpler properties. If this were the case, then everything in the universe should be conscious. This idea is thousands of years old and is called *panpsychism*.

Panpsychism, however, has never been taken seriously by science because it can hardly be falsified, and also suffers from the “combination problem” that no one has ever solved. In fact, there doesn’t seem to be any connection between what we feel and the external world. In short, if for every physical action there is an explanation that does not require consciousness, what is consciousness good for? This is why consciousness is considered *epiphenomenal*, i.e., a phenomenon that is accompanied by another phenomenon which is the true cause of what we can measure.

Accepting panpsychism implies that inner reality has a direct impact on outer reality, a possibility that the determinism of classical physics denies. No free will is possible in a deterministic universe. Consequently, our inner reality cannot have any causal power. This is equivalent to saying that the inner world is completely illusory, meaning that inner reality can only be influenced by outer reality, but not vice versa. Therefore, meaning cannot be ontological in either human beings or computers. But then, what is the reason for the existence of consciousness?

It is also peculiar that I know within myself that I am conscious, but I cannot prove it because my inner world is private and cannot be observed from the outside. I could be a zombie claiming to be conscious, and no measurement could ever prove otherwise. In fact, I can only know the feelings that another person feels only if he reveals them to me, but even in that case my knowledge cannot be certain, because that person may be mistaken or lie to me.

If we now turn our attention to free will, we realize that it crucially depends on self-consciousness, i.e., on being aware that the experience I am having is "my" experience and therefore I can direct it with "my" free will. Free will would not make sense without consciousness and without a sense of self and a sense of having an intention and purpose that can be freely pursued.

It is obvious that no deterministic system like a machine or a robot can have free will since the next state of such a system is *entirely* determined by the present state through an algorithm. Therefore, the conscious comprehension that helps us decide which path to take would not be useful because the system decides based on an algorithm, not on our state of consciousness. This is exactly what most neuroscientists are saying when they claim that consciousness is epiphenomenal.

We will soon see that the crucial properties of consciousness and free will that we sense are real, can only be held by a quantum system, and are not possible for a classical system. If we were classical systems, as most physicists think, we could not have free will, and without free will consciousness would have to be epiphenomenal and insignificant, as they describe. I claim that the deep reason we are conscious and have free will is because we are quantum entities controlling a quantum-classical body that can in turn interact with a classical deterministic world. The real me is controlling a body through which I, the quantum system, can affect the classical world with my free will. Therefore, the classical world ceases to be fully deterministic because the actions of my body in the world are caused by my non-algorithmic free will. In other words, our free will and consciousnesses *give life* to the classical universe.

3 Living Organisms vs. Robots

The most notable difference between a computer and a cell is that a computer is made of *permanent* classical matter, while a cell is made of dynamic quantum "matter". Almost all the atoms and molecules in a microchip are those that were present when the chip was first manufactured. These atoms form uncorrelated, stable physical structures that carry information in the form of electrical signals, working a bit like the roads and the traffic lights that control the movement of vehicles.

Inside a cell, on the other hand, the atoms and molecules that enter flow through it, transform, and exit so that those that leave are no longer the ones that entered it. This strategy is indispensable for carrying out the metabolic and reproductive functions of cells. Moreover, these atoms simultaneously transport information, energy, and matter to create the static and dynamic structures that allow any cell to live and experience its own existence.

The desire to explain reductionistically the holism of cells has prompted many researchers to attribute miraculous powers to the modest self-organization observed in nature in the so-called *emergent* natural phenomena. Self-organization is a spontaneous process in which an initially disordered system creates order through local interactions, as long as there is sufficient energy available. Examples of self-organization are snowflakes (the crystallization of water), certain chemical oscillations, and heat convection cells in a fluid. However, the complexity of these structures is infinitesimal compared to the complexity of a cell, just like a quantum computer with a billion entangled qubits (which does not yet exist though it may be realized in a distant future) cannot match a computer with a billion transistors.

If we compare the behavior of a paramecium with that of a tiny robot with a microchip brain, we can immediately recognize at least seven monumental differences:

1. A robot is a classical, reductionistic, and permanent organization of separable parts assembled by external agents. A cell is a quantum and classical organization of individual atoms and molecules that is holistic, dynamic, and self-reproducing in which matter, energy, and information continuously move in and out of its porous and dynamic membrane. The "hardware" of a cell is never the same, instant after instant. This difference is enormous, because the components of a cell are elementary particles, atoms, and molecules that behave primarily as described by quantum physics. Therefore, the information processing can be both quantum and classical in ways we do not yet understand given that cells have been primarily studied as biochemical classical systems. The more fundamental functioning of cells should be described as

quantum information that transforms into classical information and vice versa.

2. Robots are deterministic, with the exception of input signals which may originate from the free will of conscious agents interacting with them. A cell is inherently deterministic and indeterministic. The indeterminism derives from its quantum nature and from the creativity of its quantum consciousness and free will that direct the quantum-classical cell as a unit.
3. Robots are not autonomous and require constant supervision, unless they perform simple functions. Cells are autonomous and capable of dealing with unpredictable situations, even in hostile environments.
4. Each cell of a multicellular organism contains the entire "blueprint" of the whole organism. This incredible property does not exist in a robot, for each separable part of the robot knows nothing about the entire robot.
5. In a robot, information processing and communications are essentially digital while the input-output functions may be both digital and analog. In a cell the information processing is indivisible from the biochemical reactions that release or consume energy and construct or dismantle molecular structures, for each particle or atom is simultaneously *and* inseparably energy, matter, and information. We know very little about how to perform this type of intertwined processing, for in our computers the hardware, software, and power supply are neatly separated following the reductionism that characterizes our science and technology.
6. Robots are objective, classical systems *without* any consciousness and free will. Cells are both objective and subjective and they have consciousness and free will, even though their behaviors are strongly conditioned by the laws of quantum physics.
7. A cell has all the characteristics of a classical machine, plus many quantum characteristics presently unknown, including consciousness and free will. A cell is controllable by a quantum whole that is incommensurable with the sum of its quantum and classical parts. In a robot there is not a *whole* independent from its parts, but only the sum of its classical parts.

In summary, the crucial properties that differentiate living organisms from robots equipped with sophisticated artificial intelligence derive from the quantum nature of the cells that has been completely averaged away in the hardware of the robots.

4 Quantum Information-Based Panpsychism

My awakening experience described in the introduction irreversibly changed my perspective and redirected the course of my scientific research toward the study of consciousness. Through many other extraordinary experiences of consciousness and much reflection, I gradually developed a new model of reality consistent with both physics and my experiences. In this model, reality must have both a semantic and a symbolic aspect, i.e., information cannot be separated from its meaning as Shannon conceived. In other words, *meaning and symbols are indivisible aspects of a more fundamental reality*.

I also felt that the solution to the so-called hard problem of consciousness was hiding within quantum physics. Therefore, I began to explore quantum information and theories in which physical reality could be derived from quantum information. This is how I came across the *operational probabilistic theory* (OPT) that G.M. D'Ariano and his collaborators have developed in the last 20 years (D'Ariano 2016, 97–128). OPT showed that quantum physics could logically derive from quantum information, therefore quantum information was senior to quantum physics and the ultimate physical reality was quantum information. That was precisely the jigsaw piece I was looking for!

However, in my model information was only the symbolic aspect of reality that could not be separated from its deeper semantic aspect, i.e., from its meaning. If consciousness and free will are fundamental, information without meaning makes no sense at all, and this essential aspect was missing from D'Ariano's theory. In my model, *conscious experience* had to be another aspect of quantum information. After discussing together this point for a few years, D'Ariano finally realized that qualia could be interpreted as "the experience of a quantum system that is in a pure quantum state". We finally had found the *precise* connection between my intuitive model, based on actual experiences of consciousness, and quantum information, of which D'Ariano is a world expert.

This theory is called Quantum Information-Based Panpsychism (QIP) and has been recently published (D'Ariano and Faggin 2022, 145–191). A pure quantum state is a well-defined state that is not clonable (this is a theorem of quantum physics) and therefore it has all the extraordinary properties of a conscious experience which is also *definite* and strictly *private*. Thus, a pure state is only knowable "from the inside" by the system that is in that state, exactly like it is with our experience. On the other hand, when a quantum system interacts with a measuring apparatus, we can only observe an event describable by classical information, i.e., information that can be shared with other observers "from the outside". Therefore, classical information, which is objective and shareable, represents the *outer* physical world, whereas quantum information represents *inner* conscious experiences that are strictly private.

Note that to properly describe consciousness and free will with adequate mathematical concepts requires the extraordinary and disconcerting properties of quantum entanglement. There is no space in this paper to explain quantum entanglement save to say that it has brought to light the existence of nonlocal phenomena at the limit of human logic, without however producing contradictions. It took 80 years of theoretical and experimental efforts to understand entanglement, though it still remains suspended in an uneasy conceptual fog that no one has yet been able to completely dissipate.

Classical systems like computers use statistical properties of collections of atoms and molecules that in first approximation behave deterministically. Therefore they cannot have free will or be conscious. Note however that the determinism of a computer is only valid for as long as its circuits can faithfully recognize the two states of a bit. If the temperature were too high, for example, this recognition would fail and the computer would stop working. Moreover, if classical information were suitable to describe a conscious experience, it could be copied into the memory of a computer and thus the experience would cease to be private, which is its defining property. The clear consequence of this theory is that classical computers can neither be conscious not have free will.

It is important to realize that the representation of a conscious experience as a pure state cannot *describe* the experience itself. If it did, a pure state would be reproducible. Said differently, non-clonability expresses the existence of something private, inviolate, that can only be known by the system itself and by no one else. Prior to this theory, the existence of non-clonability could not find any reasonable explanation. The American physicist Richard Feynman would have said that "nature is absurd". Instead this "absurdity" was concealing a mystery, *the existence of interiority in the universe* that no physicist had yet realized given the materialist bias.

In other words, although I can *represent* the love I feel for my children with a pure quantum state, i.e., with a mathematical symbol, the meaning of that symbol can only be known by me. In other terms, that symbol cannot *describe* what I feel since its meaning cannot be copied, and yet the symbol can be copied. This fact makes a pure quantum state homomorphic to the qualia I feel, for I too cannot fully describe my love without violating the no-cloning theorem. I can only partially express my love with words, or with other classical symbols that are public, i.e., shareable, but that partial description cannot be confused with the love I feel within myself. Moreover, for the symbolic description to be understood by the receiver, she must already *comprehend* those symbols, which in turn requires having had a similar conscious experience than the sender.

Therefore, a pure quantum state is the mathematical entity most suitable to represent an experience because it has the extraordinary property to be *unique* and *irreproducible*, yet knowable by the system by "feeling its own state". Thus, quantum physics, which supervenes on quantum information, indicates that the ultimate ontology resides in the inner private experiences represented by quantum information. This is as far as mathematics can go to describe reality. True knowing is non-algorithmic, and can only be partially communicated with classical symbols, i.e. with algorithms. Said differently, conscious experiences are non-algorithmic because they cannot be communicated via algorithms (symbols). True knowing cannot come from the outside but can only emerge as the meaning of a conscious experience from within.

This is not difficult to understand because we have always known that the true love we feel is infinitely richer than any spoken, sung, painted, or written sentence can possibly convey.

5 Concluding Remarks

If we start from free will and consciousness as irreducible and foundational properties of nature, the entire scientific conception of reality is overturned. In this new vision, the emotional and intuitive parts of life – ignored if not belittled by materialism – return to play a central role together with the rationality that has taken us this far. We cannot let physicalism and reductionism alone define the evolution of the universe, for they do not acknowledge the free will and the consciousness that crucially determined its evolution and will increasingly do so.

Physicalism and reductionism do not recognize the holism, the open dynamism, and the self-aware consciousness of the universe that has always guided its evolution. They are adequate premises for describing the local, mechanical, and informational-symbolic aspects of classical reality, but they will hide what is most important and vital to the comprehension of the essence of the universe. If we continue to insist that these assumptions describe all of reality, we will not look into what distinguishes us from our machines, eliminating from our scientific exploration what matters most: our consciousness, our freedom, and our humanity. If we take this path, our suffering will dramatically increase because suffering represents the measure of how much we have strayed from the truth.

If, on the other hand, we take seriously the existence of the inner semantic world and begin to investigate it with love, courage, and determination, we will discover a new *Weltanschauung* that promises a creative and cooperative future, full of satisfactions and fulfillments. Life and existence cannot be defined only by mere biological and material aspects, but, above all, by the triumph of the spiritual nature of the universe that silently guides us because it is also the essence of who we are.

To explain how the universe can have created free will, life, and consciousness, the universe must have always been free, alive, and conscious, for inanimate matter can only emerge from something greater than itself. If we take this hypothesis seriously, the entire conception of reality will be transformed with enormous practical and spiritual consequences that will lead us to a brilliant future.

Humanity has been able to dramatically augment its limited physical power by understanding and utilizing the mechanical, electromagnetic, and nuclear forces far beyond the reach of human muscles. It is about to do the same with artificial intelligence, by bringing *mechanical* intelligence far beyond the reach of human brains, but not beyond the reach of what makes us human. It is time to recognize and greatly develop the lesser known powers of the human spirit that science has simply ignored.

Humanity needs to open their hearts and develop the empathy and wisdom needed to constructively use the enormous physical and informational resources at our disposal for the good of all life on our home planet. We have no choice but to raise up to the extraordinary challenges that we have created for ourselves.

References

D'Ariano, Giacomo Mauro. 2016. "Physics without Physics: The Power of Information-Theoretical Principles". *International Journal of Theoretical Physics* 56(1): 97–128. https://doi.org/10.1007/s10773-016-3172-y.

D'Ariano, Giacomo M., and Federico Faggin. 2022. "Hard Problem and Free Will: An Information-Theoretical Approach". In *Artificial Intelligence Versus Natural Intelligence*, edited by F. Scardigli. Pages 145–192. New York: Springer. http://dx.doi.org/10.1007/978-3-030-85480-5_5.

Faggin, Federico. 2021. *Silicon: From the Invention of the Microprocessor to the New Science of Consciousness*. Cardiff, CA: Waterside.

Giuseppe Vitiello

Brain, Mind, the Arrow of Time and Consciousness

Abstract: This is a short review of some features of brain modeling in quantum field theory. The brain is continuously interacting with its environment and its functional activity has to be therefore described by a dissipative dynamics. In the model, memory states are generated by the mechanism of spontaneous breakdown of symmetry. From the dissipative dynamics several consequences can be derived, such as the description of the environment in terms of the time-reversed copy of the brain, called its Double. Formally this is achieved by the procedure of doubling the degrees of freedom and the resulting state is a two-mode entangled coherent state. The act of consciousness is proposed to reside in the dialog between the brain and its Double, occurring always in the present, the "Now", where the brain and its Double indeed meet, on the surface of the time-mirror.

1 Introduction

The functional activity of the brain is characterized by its ability of focusing on a specific task for quite long time, with a time scale of the order of minutes, although the time scale of the neural activity is extremely short – of the order of milliseconds. The brain functional activity cannot be obtained by summing up, in a mechanistic way, the biochemical activity at the cellular level. This was already observed by Schrödinger, when, considering the general case of biological systems, he remarked that "it needs no poetical imagination but only clear and sober scientific reflection to recognize that we are here obviously faced with events whose regular and lawful unfolding is guided by a 'mechanism' entirely different from the 'probability mechanism' of physics" (Schrödinger 1967 [1944], 79).

What it is needed is to supplement the kinematical level of the biochemical activity by the dynamics out of which the molecular activity emerges and is by it ruled. The randomness of the chemical reactions in the environment of many different molecular species at the non-zero temperature makes very low the probabil-

Acknowledgments: To the organizers, in particular to Professors Ines Testoni and Fabio Scardigli, warm thanks for the kind invitation to the EST Conference – Eternity between Space and Time – Padua, 19–21 May 2022.

https://doi.org/10.1515/9783111313610-013

ity of the interaction between specific molecular species. This is indeed contrasting with the observed high efficiency and quite stable (compared to the scale of the chemical reactions) specialized functional activity of the brain. The available dynamics underlying the molecular activity is a quantum dynamics. Again, Schrödinger's remarks well apply to the brain when he observes that one cannot explain the "enigmatic biological stability" (Schrödinger 1967 [1944], 47) of living matter in terms of "regularities only in the average" (Schrödinger 1967 [1944], 78) originating from the "statistical mechanisms", which would be the "classical physicist's expectation" that "far from being trivial, is wrong" (Schrödinger 1967 [1944], 19). Ricciardi and Umezawa (Ricciardi and Umezawa 1967) were in fact suggesting that, in the attempt to give a dynamical description of the brain, a quantum model could satisfy the essential requirements of the observed functioning of the brain.

One further puzzle difficult to solve is the observed readiness in the brain functioning, by which at even a quite small input the brain response is almost immediate, shifting from its ongoing task to another one triggered by that small input, provided it falls within the "intentional" interests of the brain in that particular circumstances.

In the following Section 2 a brief qualitative summary of the dissipative quantum model of the brain is presented. It is discussed how the phenomenon of coherence at the microscopic molecular and neuronal level allows the transition to the macroscopic functional activities of the brain, and thus the transition between the different respective time-scales. The mathematical formalism is reported in the cited literature. In Section 3, a brief discussion of consciousness is presented and it is related to the notion of "the present" in the flowing of time. Comments concerning the entanglement relation between the brain and its environment are also discussed. Section 4 is devoted to concluding remarks. Extended reviews of the subjects discussed in the present report can be found in (Vitiello 2015; Vitiello 2022). See also Atmanspacher 2018; Atmanspacher 2020.

2 The Dissipative Quantum Model of Brain

Ricciardi and Umezawa (RU) proposed in 1967 a model of the brain based on the mechanism of spontaneous breakdown of symmetry (SBS) in many-body physics (Ricciardi and Umezawa 1967; Stuart, Takahashi, and Umezawa 1978; Stuart, Takahashi, and Umezawa 1979; Sivakami and Srinivasan 1983). Before discussing details of such a mechanism, one important remark by Karl Lashley needs to be reported. On the basis of laboratory observations, Lashley noticed that a dilemma arises:

> Nerve impulses are transmitted [. . .] from cell to cell through defined cell connections. Yet all behavior seems to be determined by masses of excitations [. . .] within general fields of activity, without reference to particular nerve cells [. . .]. What kind of nervous organization can ever account for patterns of excitations without well-defined and specialized channels of cellular communication? The problem is almost universal in the activity of the nervous system (Lashley 1942, 302–306).

Ricciardi and Umezawa were indeed observing that

> [. . .] in the case of natural brains, it might be pure optimism to hope to determine the numerical values for the coupling coefficients and the thresholds of all neurons by means of anatomical or physiological methods. [. . .] First of all, at which level should the brain be studied and described? In other words, is it essential to know the behavior in time of any single neuron in order to understand the behavior of natural brains? Probably the answer is negative. The behavior of any single neuron should not be significant for the functioning of the whole brain, otherwise a higher and higher degree of malfunctioning should be observed, unless to assume the existence of "special" neurons, characterized by an exceptionally long half life: or to postulate a huge redundancy in the circuitry of the brain. However, to our knowledge, there has been no evidence which shows the existence of such "special" neurons, and to invoke the redundancy is not the best way to answer the question (Ricciardi and Umezawa 2004, 256).

Thus they proposed a model based on the formalism of SBS, which can occur in quantum field theory (QFT) due to the existence of physically inequivalent representations (Hilbert spaces) of the canonical commutation relations (CCR), each representation describing indeed a different dynamical regime, i.e. with physically different contents, of the system under study (Umezawa 1993, 52; Blasone, Jizba, and Vitiello 2011, 58–66). In quantum mechanics (QM), the Stone-Von-Neuman theorem states that all the representations of the CCR are physically equivalent. The brain activity is characterized by a flow of transitions through different dynamical regimes (Vitiello 1995; Vitiello 2001; Freeman and Vitiello 2006; Freeman and Vitiello 2008; Freeman, Livi, Obinata, and Vitiello 2012) and therefore QFT, not QM, provides the convenient mathematical framework for brain modeling.

SBS occurs when the system vacuum (i.e. the ground state of one of the infinitely many inequivalent representations) is not invariant under the full symmetry group of the dynamics defined by the field equations. Symmetry is thus broken in the system vacuum state. The symmetry breaking is not due to a modification of the field equations (in that case one says that explicit BS occurs), but to the action of a triggering input acting on the system, which then "spontaneously" sets in one of the available ground states.

In the RU model the symmetry breaking trigger is assumed to be the input through which information from the external world (including the subject body) reaches the brain through the perception channels. The crucial point is that in

QFT the Goldstone theorem holds, stating that when SBS occurs then long-range phase correlations are dynamically generated in the vacuum (Goldstone, Salam, and Weinberg 1962; Umezawa 1993, 66; Blasone, Jizba, and Vitiello 2011, 112). This is an experimentally well established result in many-body physics. Long-range is referred to the typical scale of the system components (the molecular scale in the RU model).

A net of correlations is thus induced in the vacuum state, establishing in it an ordered structure. In QFT, order is therefore dynamically generated by the SBS. Order is lack of symmetry. In a quantum theory, one may associate to the correlation waves a quantum, called in the SBS case the Nambu-Goldstone (NG) quantum or mode, which is massless in the infinite volume limit; otherwise, in the presence of boundaries or impurities, it may acquire a non-zero effective mass. One may then refer to the correlation net in the vacuum also in terms of coherent condensation of the NG quanta. *Coherent* condensation means that the NG quanta share the same phase, they are phase-correlated and therefore describe a collective mode which extends over the whole volume where SBS occurs. Coherent correlations are those that coexist without negative interferences. Coherence is therefore the key mechanism by which a collection of elementary individual components manifests at a macroscopic level a *collective behavior*. Coherence contributes to tame the randomness of the chemical activity, establishing the observed long-range correlations among molecules (Lechelon et al. 2022), thus providing the dynamical transition from the microscopic level to the macroscopic one. Familiar examples or coherent systems are crystals, magnets, and in general ordered structures. They are long lived systems provided temperature is below a critical threshold T_C, e.g. the Curie temperature for magnets. The range of critical temperature is quite large, from very low temperatures, as e.g. for compounds of niobium superconductors ($T_C = -252$), to high temperature for crystals (the diamond loses its coherent (crystal) ordering, i.e. it melts, at about 3545 °C at atmospheric pressure). In the crystals, the NG quanta of the elastic waves, responsible of the crystal ordering, are the phonons. In the magnets, the NG quanta of the spin waves responsible of the magnetic ordering are the magnons. Notice that QFT coherence is not affected by the QM decoherence phenomenon. QFT coherence is of different nature than the one considered in QM. C*oherence* dynamically generated in QFT plays a crucial role in particle physics and condensed matter (Umezawa 1993, 16; Blasone, Jizba, and Vitiello 2011, 37; Srinivasan 1976).

The correlation net represents, according to the RU model, the memory, i.e. the recording of the information carried by SBS triggering input. The vacuum so structured is then the "memory state". The recalling of such a memory is generated by exciting out of the memory state the corresponding NG mode (the correlation mode) induced by an input similar, not necessarily identical, to the original SBS triggering input.

More details are not reported here for brevity. However, a difficulty of the many-body model needs to be mentioned: once a memory has been recorded, the recording of a memory related to a successive input will erase the previously recorded memory. In order to avoid such a difficulty, the RU model has been extended (Vitiello 1995; Vitiello 2001) to the "dissipative quantum model of brain", so to include the dissipative character of the brain dynamics.

In brain modeling one cannot avoid to consider that the brain is a system permanently *open* on the world surrounding it, it is never a closed system. Therefore its dynamics is a dissipative dynamics. Here the word dissipative (and dissipation) means that the brain releases, but also receives energy in its many different forms from the environment. In the dissipative model, the QFT dissipation formalism is adopted, where the dissipative system evolution is described as a path or trajectory over the manifold of inequivalent representations whose vacuum states are labeled by the time parameter. Thus, depending on the time at which different triggering inputs occur, different trajectories of different initial time (for different memories) are generated. The physical inequivalence (unitary inequivalence in the QFT jargon) of the representations at different times for different memories guaranties that intersections among different trajectories do not occur, neither a trajectory intersects itself, and therefore "confusion" of memories cannot occur (although, boundary effects may sometimes cause accidental intersection). In the dissipative model, the memory capacity is thus very large since infinitely many inequivalent representations are available and different coherent condensate structures of the NG modes are realized in different vacua (different memory states), the difference consisting in the different spectral structure of the condensed NG modes (Vitiello 1995).

Once a memory has been recorded by a subject as a result of its perceptive experience (input), then "*Now you know it!*" can be said to the subject, meaning that memory recording produces a partition on the time axis, introducing the distinction between the time "before" recording, i.e. the past, and the time "after" it, i.e. the future. The past can be now distinguished from the future. Memory recording breaks time translational symmetry and time reversal symmetry. The *arrow of time* is thus generated describing the succession of time (time ordering) and its irreversible evolving direction. Moreover, the succession of time so introduced also produces the perception of causal succession of events generated by the subject actions. The *action-perception* cycle is then recognized to be a distinctive feature of the brain functional activity, characterizing the *to-be-in-the-world* of the subject (Freeman and Vitiello 2016).

A final remark concerns which symmetry gets broken under the triggering input. In the RU model, as well as in the dissipative quantum model, such a symmetry is the rotational symmetry of the electrical dipoles characterizing the constituent molecules of the brain and its neuronal cell (Jibu and Yasue 1995, 152; Vitiello 1995). Water molecules are much lighter than the other biomolecules and in number they are about the 90% of the total of present molecules. The SBS of the electrical molecular dipoles thus generates long-range dipole waves, and the associated NG modes are dipole wave quanta (DWQ) (Del Giudice, Doglia, Milani, and Vitiello 1985; Del Giudice, Doglia, Milani, and Vitiello 1986). The dipole wave correlation net, i.e. in terms of quanta, the DWQ coherent condensate constitutes the dynamically generated background promoting neuronal dynamical interaction. The correlation net provides a sort of wave guide (ordered pattern) promoting specific biochemical interactions, thus promoting the high biochemical efficiency against the otherwise fully random chemical activity, which constitutes the fatal obstacle unsurmounted by neuroscience approaches solely based on biochemistry. It has to be stressed that the RU model and the dissipative quantum model absolutely do not exclude the necessity of the most accurate biochemical analysis; they state, however, that it is by itself not sufficient. The QFT models explain on a dynamical basis the observed high efficiency of the biochemical activity, otherwise inexplicable on the basis of purely random chemist (recall the above reported Schrödinger's remarks on the "enigmatic biological stability").

The dynamical generation of the ordered net of long-range correlations also provides a hint to a possible resolution of the above-mentioned Lashley's dilemma. The collective modes of DWQ might shed some light on the nature of Lashley's "general fields of activity, without reference to particular nerve cells", and the dynamical coherent correlation generated by the dipole waves might indeed accounts for the observed "patterns of excitations without well-defined and specialized channels of cellular communication", also promoting the observed ephapsis phenomena (touching, non-synaptic neuron-to-neuron interaction) (Grundfest 1959). Remarkably, the short life time of coherent long-range correlation among water molecular dipoles is favorable to the system fast recovery from specific ordered pattern, ready to undergo a "phase transition" to a differently ordered pattern. Such a property is crucial to avoid functional rigidity and it allows the observed coexistence, otherwise contradictory, of functional stability and, at once, functional readiness to respond to rapidly changeable environmental boundary conditions. More comments on this aspect crucial to the survival and the functional efficiency of the system will be presented in the following.

3 Consciousness. The Dialog with the Double

The dissipative character of the brain dynamics introduces us to a special feature of the brain functional activity, namely to the permanent, unavoidable dialog of the brain with the world *as perceived* by the subject. One specific mathematical requirement to be satisfied in the description of quantum dissipation in QFT is the balance of the flows of energy between the dissipative system and the environment in which it is embedded, so that the couple {system, environment} constitutes a closed system: the amount of energy in-going into the system has to be exactly equal to the one out-going from its environment, and vice-versa. Since out-going flows may be described by reversing the time direction adopted for the in-going flows (and vice-versa), in a schematized way, from the standpoint of the energy in-out-flows balancing, the environment may be described as the time-reversed image of the system (or vice-versa) (Vitiello 1995). Assuming that A_k describes some degree of freedom of the system, its time-reversed image, denote it by $\tilde{A}_k$, will describe the environment. The dissipative system is therefore described in terms of the couple $\{A_k, \tilde{A}_k\}$. In this construction, the environment, being the exact time-reversed copy of the system, is called the system *Double*. The permanent dialog between the brain and its environment is therefore depicted as the dialog between the brain and its Double. In the dissipative model the act of consciousness resides in such a dialog. There the identity of the two actors, becoming mutually dependent and dynamically evolving in time, gets continuously shaped. Since time evolution is described by trajectories going through unitary inequivalent, i.e. physically different, representations, each one labeled by each instant of the time *t*, like a photogram in a flowing movie, the act of consciousness lives only in *the present*, on the surface of the time-mirror where the self and its Double meet, in the *Now*. Consciousness has thus the magic flavor of the halting the flow of time (an image of eternity? achieving a non-dissipative state? the monadic vision of Globus (Globus 2003). As observed elsewhere (Vitiello 2004a, 330), perhaps,

> in these Nows is realized the primary property of consciousness, the one of self-questioning (Desideri 2003, 21–28), i.e. the unveiling the Double, and the photographer's "sur-prise" . . . "when at the precise instant an image suddenly stands out and the eye stops" forcing "the time to stop his course" (Prete 2003, 4): "and suddenly, all at once, the veil is torn away, I have understood, I have *seen*" (Sartre 1990, 171).

Unveiling the Double is then to see and to be seen, the *συνειδώς*, the being conscious of the ancient Greeks, which literally is to "see together", indeed; or, as in the lifting the veil in the Prete's photo-objects, "more precisely, to have a perception of this togetherness as a whole and to understand that it was made of two images in strong relation" (Prete 2003, 4); or else Bohm's self-recursive mirroring loops of the spontaneous and unrestricted act of "lifting into attention" (Bohm 1980, 35): *συνειδώς* then comes to be confidants, secret friends (Bandini 2002, 147), to be each other "witness".

In the dissipative quantum model, the "undivided" and "undividable" relation with the Double is described by the entanglement of the $\{A_{\mathbf{k}}, \tilde{A}_{\mathbf{k}}\}$ modes in the brain coherent state at each time t in its evolution (Vitiello 2004a, 322; Pessa and Vitiello 2004; Sabbadini and Vitiello 2019). From a mathematical standpoint, the entanglement is a direct consequence of the dissipative character of the dynamics and of the coherence of the condensate. Moreover, at each time t the brain coherent state minimizes the free energy and it plays the role of an "attractor" for the system dynamical evolution. The set of inequivalent vacua states constitute then what in the language of nonlinear dynamical systems is called the landscape of attractors. The space of coherent states form a Khälerian manifold (Perelomov 1986, 185–203), and one can show that the trajectories going through the manifold of inequivalent representations (in a succession of phase transitions) are classical chaotic trajectories (Vitiello 2004b; Pessa and Vitiello 2004). Observations by Freeman (Freeman 1991; Freeman 2004), by Kozma and Freeman (Kozma and Freeman 2002) and Skarda and Freeman (Skarda and Freeman 1987), also confirmed in successive observations in neuroscience (Plenz and Thiagaran 2007; Petermann, Thiagarajan, Lebedev, Nicoleli, Chialvo, and Plenz 2009), have shown that chaotic dynamics plays indeed a relevant role in brain functional activity (Tsuda 2001; Freeman and Vitiello 2006). Since in a chaotic regime, small changes in the initial conditions induce diverging trajectories, the chaotic dynamics makes possible the readiness of brain responses to rapid changes in the surrounding environment, an aspect already mentioned above. This shows that a noisy environment, although being source of some higher or lower disturbing effects, however it may also introduce some element of interest to the attentive brain, that even if engaged, along a certain trajectory, in pursuing a specific task, might be searching for a novelty, something not directly or immediately related to that path or trajectory, thus causing an even slight shift in the initial conditions with consequent switching, due to chaoticity, to a diverging trajectory. Such kind of "erratic" regime might even be the source of a "novel vision", a new, i.e. not necessarily, not logically implied by previous steps, solution to a problem. Perhaps, this is the source of the "intuition", the *intueri*, that *instantaneous apprehension* occurring "without the conscious use of reasoning" (see "intuition" in the Webster's Dictionary, 1968). Perhaps, *pensare* (to think) has its real root in the *errare* (to wander) (Minati and Vitiello 2006; Vitiello 2022).

4 Concluding Remarks

As discussed above, the brain in constructing an image of the environment, constructs the time-mirrored image of itself, its Double. Thus, "brains imagine hypotheses about the world . . . mirroring them in time" (Freeman and Vitiello 2016, 7). In this process, it is the Double, that exploring backward in time the "space of the memories", formulates the hypotheses from new experiences and updates the landscape of the attractors, formed on the basis of previous experiences, according to the best matching of the new hypotheses with one of the attractors, or creating a new attractor if such a matching fails. It has been proposed (Freeman and Vitiello 2016, 16) that this is the *mind* activity, the one of the Double, while the brain (*matter*) is responsible for the (new) consequent actions. The hypotheses thus screened and acquired and the actions on them based, with consequent new perceptive experiences and new hypotheses, in the action-perception cycle, construct the *knowledge*, aimed to the best *to-be-in-the-world* of the subject. This sequence of steps provides the *meaningfulness* of the hypotheses and of the actions within the world context in which the brain is embedded. The formal analysis of such a process has led to the definition of the pragmatic information index (Freeman and Vitiello 2006, 99) after Atmanspacher and Scheingraber (Atmanspacher and Scheingraber 1990). In the model, mind activity and brain activity do not have separate individual existence, they are formally described as entangled modes.

In this review, visual experiences during dreams, meditation and non-ordinary brain activity states (Globus 2019; Re and Vitiello 2020a; Re and Vitiello 2020b) have not been considered for brevity. In such experiences, occurring under conditions of very few and weak links between the brain and the world, the consequential succession in time evolution is violated since the perception of the distinction between the brain and the Double is much lowered or even lost, so that "fragments" and "bits" of memories are not assembled according to their original time sequence in the awake state.

References

Atmanspacher, Harald. 2018. "Walter Freeman. I Did It My Way". *J. Consciousness Studies* 5(1–2): 39–44.

Atmanspacher, Harald. 2020. "Quantum Approaches to Consciousness". *The Stanford Encyclopedia of Philosophy* (Summer 2020 Edition), edited by E.N. Zalta. https://plato.stanford.edu/archives/sum2020/entries/qt-consciousness/, last accessed 18 October 2022.

Atmanspacher, Harald, and Herbert Scheingraber. 1990. "Pragmatic Information and Dynamical Instabilities in a Multimode Continuous-Wave Dye Laser". *Can. J. Phys* 68: 728–737. DOI: 10.1139/p90-105.

Bandin, Fernando 2002. "La Coscienza e il Tempo". In *Exploring Consciousness*, edited by K.E. White, 147–153. Milan: Fondazione Carlo Erba.

Blasone, Massimo, Peter Jizba, and Giuseppe Vitiello. 2011. *Quantum Field Theory and Its Macroscopic Manifestations*. London: Imperial College Press.

Bohm, David. 1980. *Wholeness and the Implicate Order*. London: Routledge and Kegan Paul.

Del Giudice, Emilio, Silvia Doglia, Marziale Milani, and Giuseppe Vitiello. 1985. "A Quantum Field Theoretical Approach to the Collective Behavior of Biological Systems". *Nucl. Phys. B* 251(FS 13): 375–400. DOI: 10.1016/0550-3213(85)90267-6.

Del Giudice, Emilio, Silvia Doglia, Marziale Milani, and Giuseppe Vitiello. 1986. "Electromagnetic Field and Spontaneous Symmetry Breakdown in Biological Matter". *Nucl. Phys. B* 275 [FS17]: 185–199. DOI: 10.1016/0550-3213(86)90595-X.

Desideri, Fabrizio. 2003. "The Self-Transcendence of Consciousness towards Its Models". In *Brain and Being*, edited by Gordon G. Globus, Karl H. Pribram, and Giuseppe Vitiello, 21–28. Amsterdam: John Benjamins.

Freeman, Walter J. 1991. "The Physiology of Perception". *Scientific American* 264: 78–87.

Freeman, Walter J. 2004. "How and Why Brains Create Meaning from Sensory Information". *Int. J. Bifurc.Chaos* 14: 513–530. DOI: 10.1142/S0218127404009405.

Freeman, Walter J., Roberto Livi, Masashi Obinata, and Giuseppe Vitiello. 2012. "Cortical Phase Transitions, Non-Equilibrium Thermodynamics and the Time-Dependent Ginzburg-Landau Equation". *Int. J.Mod. Phys. B* 26: 1250035. DOI: 10.48550/arXiv.1110.3677.

Freeman, Walter J., and Giuseppe Vitiello. 2006. "Nonlinear Brain Dynamics as Macroscopic Manifestation of Underlying Many-Body Dynamics". *Physics of Life Reviews* 3: 93–117. DOI: 10.48550/arXiv.q-bio/0511037.

Freeman, Walter J., and Giuseppe Vitiello. 2008. "Dissipation, Spontaneous Breakdown of Symmetry and Brain Dynamics". *J. Phys. A: Mathematical Theory* 41: 304042. DOI: 10.48550/arXiv.q-bio/0701053.

Freeman, Walter J., and Giuseppe Vitiello. 2016. "Matter and Mind Are Entangled in Two Streams of Images Guiding Behavior and Informing the Subject through Awareness". *Mind and Matter* 14: 7–24.

Globus, Gordon. 2003. *Quantum Closures and Disclosures:Thinking-Togheter Postphenomenology and Quantum Brain Dynamics*. Amsterdam: John Benjamins.

Globus, Gordon. 2019. "Lucid Existenz during Dreaming". *Int J Dream Res*. 12: 70–74.

Goldstone, Jeffrey, Abdus Salam, and Steven Weinberg. 1962. "Broken Symmetries". *Phys. Rev.* 127: 965–970. DOI: 10.1103/PhysRev.127.965.

Grundfest, Harry. 1959. "Synaptic and Ephaptic Transmission". In *Handbook of Physiology*, edited by J. Fields, W. Magoun, and V. Hall, Vol. 1, 147–197. Baltimore: Williams & Wilkins.

Jibu, Mari, and Kunio Yasue. 1995. *Quantum Brain Dynamics and Consciousness*. Amsterdam: John Benjamins.

Kozma, Robert, and Walter J. Freeman. 2002. "Classification of EEG Patterns Using Nonlinear Dynamics and Identifying Chaotic Phase Transitions". *Neurocomputing* 44: 1107–1112. DOI: 10.1016/S0925-2312(02)00429-0.

Lashley, Karl S. 1942. "The Problem of Cerebral Organization in Vision". In *Biological Symposia*, Vol. 7: *Visual Mechanisms*, edited by H. Klüver, 301–322. Lancaster: Jacques Cattell Press.

Lechelon, Mathias, Yoan Meriguet, Matteo Gori, Sandra Ruffenach, Ilaria Nardecchia, Elena Floriani, Dominique Coquillat, Frédérich Teppe, Sébasyien Mailfert, Didier Marguet et al. 2022. "Experimental Evidence for Long-Distance Electrodynamic Intermolecular Forces". *Sci. Adv.* 8: eabl5855. DOI: 10.1126/sciadv.abl5855.

Minati, Gianfranco, and Giuseppe Vitiello. 2006. "Mistake Making Machines". In *Systemics of Emergence: Research and Development*, edited by G. Minati, E. Pessa, and M. Abram, 67–78. Berlin: Springer.

Perelomov, Askold. 1986. *Generalized Coherent States and Their Applications*. Berlin: Springer.

Pessa, Eliano, and Giuseppe Vitiello. 2004. "Quantum Noise Induced Entanglement and Chaos in the Dissipative Quantum Model of Brain". *Int. J. Mod. Phys.* B 18: 841–858. DOI: 10.48550/arXiv.quant-ph/0406161.

Petermann, Thomas, Tara C. Thiagarajan, Mikhail A. Lebedev, Miguel A.L. Nicoleli, Dante R. Chialvo, and Dietmar Plenz. 2009. "Spontaneous Cortical Activity in Awake Monkeys Composed of Neuronal Avalanches". *Proc. National Acad. Sci.* 106(37): 15921–15926. DOI: 10.1073/pnas.0904089106.

Plenz, Dietmar, and Tara C. Thiagaran. 2007. "The Organizing Principles of Neural Avalanches: Cell Assemblies in the Cortex". *Trends Neurosci* 30: 10110. DOI: 10.1016/j.tins.2007.01.005.

Prete, Nadia. 2003. "Doubling Image to Face the Obscenity of Photography". In *Brain and Being*, edited by G.G. Globus, K.H. Pribram, and G. Vitiello, 255–266. Amsterdam: John Benjamins.

Re, Tania, and Giuseppe Vitiello. 2020a. "Nonlinear Dynamics and Chaotic Trajectories in Brain–Mind Visual Experiences during Dreams, Meditation and Non-Ordinary Brain Activity States". *OBM Neurobiology* 4(2). DOI: 10.21926/obm.neurobiol.2002061.

Re, Tania, and Giuseppe Vitiello. 2020b. "On the Brain–Mind Visual Experiences". In *Proceedings IEEE – IJCNN, Glasgow*. DOI: 10.1109/IJCNN48605.2020.9207327.

Ricciardi, Luigi M., and Hiromi Umezawa. 1967. "Brain and Physics of Many-Body Problems". *Kybernetik* 4: 44–48.

Ricciardi, Luigi M., and Hiromi Umezawa. 2004: "Brain and Physics of Many-Body Problems". In *Brain and Being*, edited by G.G. Globus, K.H. Pribram, and G. Vitiello, 255–266. Amsterdam: John Benjamins. DOI: 10.1007/BF00292170.

Sabbadini, Shantena A., and Giuseppe Vitiello. 2019. "Entanglement and Phase-Mediated Correlations in Quantum Field Theory. Application to Brain–Mind States". *Applied Science* 9: 3203. https://doi.org/10.3390/app9153203.

Sartre, Jean P. 1990. *La nausea*. Turin: Giulio Einaudi.

Schrödinger, E. 1967 [1944]. *What Is Life?* Cambridge: Cambridge University Press.

Sivakami, S., and V. "Seenu" Srinivasan. 1983. "A Model of Memory". *J. Theor. Biol.* 102: 287–294. DOI: 10.1016/0022-5193(83)90366-1.

Skarda, Christine A., and W.J. Freeman. 1987. "How Brains Make Chaos in Order to Make Sense of the World". *Brain. Behav. Sci.* 10: 161–195. DOI: 10.1017/S0140525X00047336.

Srinivasan, V. "Seenu". 1976. "On the Connection between the Boson Transformed Ground State and the Coherent State". *Progr. Theor. Phys* 55: 939–940.

Stuart, C.I.J.M., Yukio Takahashi, and Hiromi Umezawa. 1978. "On the Stability and Non-Local Properties of Memory". *Journal of Theoretical Biology* 71: 605–618. DOI: 10.1016/0022-5193(78)90327-2.

Stuart, C.I.J.M., Yukio Takahashi, and Hiromi Umezawa. 1979. "Mixed System Brain Dynamics: Neural Memory as a Macroscopic Ordered State". *Foundations of Physics* 9: 301–327. DOI: 10.1007/BF00715185.

Tsuda, Ichiro. 2001. "Towards an Interpretation of Dynamic Neural Activity in Terms of Chaotic Dynamical Systems". *Behav Brain Sci* 24: 793–810. DOI: 10.1017/s0140525x01000097.
Umezawa, Hiromi. 1993. *Advanced Field Theory: Micro, Macro and Thermal Concepts*. New York: American Institute of Physics.
Vitiello, Giuseppe. 1995. "Dissipation and Memory Capacity in the Quantum Brain Model". *Int.J. of Mod. Phys. B* 9: 973–989. DOI: 10.48550/arXiv.quant-ph/9502006.
Vitiello, Giuseppe. 2001. *My Double Unveiled*. Amsterdam: John Benjamins.
Vitiello, Giuseppe. 2004a. "The Dissipative Brain". In *Brain and Being*, edited by G.G. Globus, K.H. Pribram, and G. Vitiello, 255–266. Amsterdam: John Benjamins.
Vitiello, Giuseppe. 2004b. "Classical Chaotic Trajectories in Quantum Field Theory". *Int. J. Mod. Phys.* B 18: 785–792. DOI: 10.48550/arXiv.hep-th/0309197.
Vitiello, Giuseppe. 2015. "The Use of Many-Body Physics and Thermodynamics to Describe the Dynamics of Rhythmic Generators in Sensory Cortices Engaged in Memory and Learning". *Current Opinion in Neurobiology* 31: 7–12. DOI: 10.1016/j.conb.2014.07.017.
Vitiello, Giuseppe. 2016. "Filling the Gap between Neuronal Activity and Macroscopic Functional Brain Behavior". In *Cognitive Phase Transitions in the Cerebral Cortex – Enhancing the Neuron Doctrine by Modeling Neural Fields*, edited by R. Kozma and W.J. Freeman, 239–249. Cham: Springer International.
Vitiello, Giuseppe. 2018. The Brain and Its Mindful Double. *Journal Consciousness Studies* 25: 151–176.
Vitiello, Giuseppe. 2022. "The Brain is Not a Stupid Star". In *Artificial Intelligence versus Natural Intelligence*, edited by F. Scardigli, 107–144. Cham: Springer Nature.
Webster's New World Dictionary. 1968. *College Edition*. Cleveland: World Publishing.

Third Part: **Eternity, Time and Faith**

Kurt Appel

The Eighth Day. Biblical Time as Openness of Chronological Time

Abstract: This paper examines how the biblical conception of time differs from, and poses a challenge to, our currently predominant chronological understanding of time. A chronologically conceived time essentially implies a physicalist understanding of the world, tracing the universe back to elementary particles and the physical forces acting upon them. Crucial here is the notion of an unbroken causal sequence of connections between entities. Leibniz and Hegel, whilst never invalidating the noetic validity of natural sciences for humankind, both criticised this view as unbalanced. They claimed that space and time cannot be understood independently of the subject. Our world is not a mechanism, but has subject-like dimensions. Similar thoughts are also expressed in the Bible. The key figure is the eschatological "eighth day", which inscribes a fundamental non-objectifiable openness into time which renders void the gapless linear causality of space-time. The eighth day, which manifests the eschatological meaning of the Sabbath, provides the basis of freedom and alterity. Importantly, it reveals itself in the time of the feast, thus transcending any functional determination of Being.

1 The Mechanistic Understanding of Time

Our current understanding of time is a chronological one. The predominant assumption is that the universe emerged from the "Big Bang" 13.7 billion years ago, with the formation of our Earth 4.6 billion years ago. Continuing this timeline, we encounter the first life in Earth's primordial oceans 3.75 billion years ago. If we imagined our universe's timespan until the present as one day, the period of the linguistically capable human being (of the last 300 000 years) would comprise only a few milliseconds.

Today's science-shaped worldview does not offer much hope for our future. According to an optimistic estimate, the resources that currently provide us the opportunity to lead lives of relative material comfort, will suffice for only a few hundred years. After this, our way of life as far as we can imagine it today will no longer be sustainable. We do not know whether a nuclear or ecological super-disaster may occur after or even before this point in time, but unfortunately such a scenario cannot be ruled out. Yet even if we allow for a prolonged existence of humankind, we will eventually encounter limits: In 800 million years, solar radia-

https://doi.org/10.1515/9783111313610-014

tion will render life on Earth impossible. In 10^{100} years, due to entropy, all structures of the cosmos will have dissolved. According to current observational analyses of cosmic expansion, light from other galaxies will no longer reach us in a few billion years. The bleak prognosis for the universe is one of disintegration and increasing desolation. Within this cosmic totality, man and his creations constitute only a minor interlude.

The question, however, is whether humankind ought to be inserted into chronological time in this way. For cosmological theories also presuppose the knowing observer. Even if man could abstract from any meaning, he cannot abstract from himself as an observer. We can describe the world physically through particles and their interactions, which function according to the law of causality. However, man cannot be absorbed in this causality. For even if he mentally were to conceptualise himself altogether causally, he would still remain the observer of this process, and thus not on the same level with it.

The problem with an understanding of time as outlined above is *firstly* that it equates Being to physical phenomena. The principal law applied when thus determining time is causality. An absolute order of time progresses from cause to cause, and nothing can escape its linear connection. Neither life nor self-consciousness should be interpreted purely causally, and also conversations elude linear causality. If there was a complete causality of our biography, the human being could not remain a human being for even a single moment. This becomes apparent when we attempt to narrate our lives: we could not create a gapless biography for even one minute, because our history contains a continual recreation involving the motivational horizons of both past and future, from which the present arises. If we imagined man embedded in a gapless causal chain that he simultaneously observes, he would immediately position himself in it and interpret this chain, i.e. recreate time.

Secondly, a purely chronological understanding of time is problematic in that the world is thought of as a quantifiable mechanism that levels out all qualities. How could we derive any substantive value from such a view? The annihilation of any specific substance corresponds to the idea of the demise of the world through the entropy of the expanding universe. The cosmos devours its own observers and dissolves into nothingness. One consequence of this would lie in the impossibility of a qualitative difference between past, present and future, which is constitutive for human reasoning (Koch 2006).

The decisive question is whether time can be grasped independently of both the substance it contains and the subject that experiences it.

2 The Time of the Biblical Creation Narrative

The qualitative understanding of time in the biblical creation narrative is opposed to this quantifying concept. Upon careful reading, its structure appears conspicuous: the creation of the world is symbolically expressed as a seven-day week. Time as a motif is emphasised in connection with days 1, 4 and 7 (Zenger 1983). It thus represents the beginning, centre, and end of the opening narrative of the Bible.

In ancient times, books were named after their first sentence. The title of the Bible is therefore "At the beginning God created the heavens and the earth". God's creating in the beginning is thus the keynote to the whole Bible. The second sentence of the Bible names the origin that precedes creation: "But the earth was desert and void". This describes the initial situation of the world from the perspective of mythologising cultures: The first reality of the world is boundless chaos, which symbolises death. The first face of time is nothingness, which afflicts gods and men, heaven and earth. Western metaphysicians since Parmenides have responded to this voidness of time by attempting to cross over from the temporal into a timeless beyond (Heinrich 1966).

The Bible approaches this differently. The account of Day One already serves this purpose. "Evening came and morning came: One day". This represents the fundamental unity of time, in which the timeless, eternal chaos is overcome. Of course, this "one day" must not be understood chronologically in the sense of 1440 minutes, nor as a stretch of time from sunrise to sunset. Rather, this day consists of *three times*, beginning with the *evening* (evening time; time I), continuing through the *night* (nocturnal time of death; time II) and ending in the *morning* (eschatological time; time III). This sequence reflects the biblical motif of the starting point of time as the evening, moving towards nightly death. Night symbolises death, as man loses consciousness in sleep and commonly dies during night. The unconscious realm of night and sleep also constitutes a withdrawal of time. The third form of time as its end point is morning time wherein night and death are overcome. Theologically, this is termed the "eschaton", which is, however, neither the continuation of time nor the return to evening time. We also encounter three-stage time in the account of Jesus' resurrection on the *third* day. The first day corresponds to world time that has lapsed into death, i.e. time I of the one-day, the second day corresponds to its interruption in the form of death (i.e. time II of the one-day), and the third day denotes the new creation of time (i.e. time III of the one-day).

Eschatological time is the transcendence of chronological time as expressed, for example, in the feast (Bahr 2008) and in the liturgy of the Christian and Jewish tradition. Biblical time does not aim at chronology, but at the feast as the deepest figure of time.

The next step is to examine what characterises the festive season, and whether there is any philosophical mediation between chronological-scientific time and the biblical festive season. The significance of the feast in the Bible is also shown by the further days of creation: the second and third days prepare the festive space and the festive decoration of the earth by the creation of plants. Just like a table is set, so Earth is adorned with plants. The fourth day, which is the centre of the week, leads into the middle of the seven-day creation: it deals with the creation of the sun and moon, insofar as these are essential for the festive calendar. In the centre of time, therefore, is the feast. Day Five and Day Six have the function of inviting guests to the universal feast. The provisional end point is reached with the human being, whose purpose is supporting God in the organisation of the feast. As God's representative, man has the function of making the festive origin and the festive meaning of the world present.

World time is completed after six days. Nevertheless, creation ends with the seventh day, transcending world time and inscribing an addition onto it, which has the effect that world time is never completely closed, never becomes self-identical. It is precisely this open addition that symbolises the transcendence of time in freedom, which shatters every worldly causality and solemnly expresses itself in the seventh day. The seventh day renders functionality obsolete. It carries its meaning within itself, and in this freedom and solemnity becomes the event of God's epiphany. Biblically, the seventh day reveals itself to the creation, represented in the human being, as a festive transcendence of the constraints of chronological time. God is not supreme causality, but rather the event of festive time, to which man responds in joy.

This understanding of time gains special expression in Christianity with the idea of the eighth day, the Sunday, within which the six days of world time and the Sabbath, i.e. the seventh day, unite. The formula of Sunday would be 6 + 1 + 1. It is, inasmuch as doubling in the Bible always expresses a special emphasis, the transcendence of the transcendence of the seventh day, its peculiarity being that it is also the first day, or rather the true one-day, that is, the true foundation of time. It follows that the world is created in feast, and that the festive addition of the seventh day coincides with each day of world time. Time thus contains both causal world time *and* its festive, free, and open "more".

The question now arises whether and how the biblical concept of time can be reconciled with a philosophical approach to time.

3 Leibniz: The Self-Temporalising Monad

Leibniz can still serve as the starting point for a philosophical critique of a purely quantitative concept of time. In his mill allegory (*Monadology*, § 17), he points out that a mechanistic view of the world is not sufficient to be able to recognise it in its concreteness:

> And if we suppose that there were a machine whose structure makes it think, feel, and have perception, we could imagine it increased in size while keeping the same proportions, so that one could enter it as one does with a mill. If we were then to go around inside it, we would see only parts pushing one another, and never anything which would explain a perception (Strickland 2014, 17 [§ 17]).

A decisive question for Leibniz is that of the unity of Being as a transcendental principle of its reality. Why can we name Being as unity and thereby elevate it into the word? The space-time machinery knows only separatedness. Therefore, when something is addressed as a unity, this unity is already presupposed. Leibniz calls this unit a "monad". Since we know no units except subjects, monads are to be thought of in analogy to a subject. Leibniz outlines a scale of such units with an increasing degree of subjectification, i.e. the accent shifts gradually from mere objecthood towards the sphere of subjectivity. In the realm of inorganic nature, unity is manifested in its capacity for differentiation as well as in its inherent gravity. One could therefore speak of "inner unity" (Klein 2006, especially 103–112). Unity, and thus subject-likeness, becomes clearer in the realm of the living, especially with higher mammals, who possess memory and can thus put the world they perceive into a temporal unity. At the end of this scale, we are dealing with unities in the sphere of the human being, who is not only aware of his own unity, but is also able, by means of reason, to conceive of the entire world as a one.

With regard to the space-time problem, it must be emphasised that inner unity, as a transcendental heuristic scheme under which the world must be considered, is entirely ideational. However, every inner entity ("monad") mediates itself spatiotemporally. In this respect, every entity is subject-object or, as Leibniz also calls it, *vis primitiva* and *vis derivativa*. According to a terminology derived from Husserl, it could be said that the monad temporalises (and spatialises) itself. In other words, no monad without time, but also no time without monad. There is no time without subject and no subject without time. When each monad temporalises and spatialises itself, there is no longer an "outside" of the monad. With the appearance of each monad, the world comes into Being. This does not mean that the monad is omnipotent or that only one monad has existed. Rather, the consequence of the monadic structure of the universe is that each monad also presupposes the time preceding its position, mirroring an infinite number of other monads, albeit at different intensities of mediation. We are, for example, connected to the Earth differently than to our parents or to the neighbouring star Alpha Centauri.

The presupposition of every monad is also the positing set by the monad itself, which does not exist without it. The child is both the consequence and the cause of its parents. I am only in relation to the sun and the sun only in relation to me. Leibniz further shows that each monad as a mirror of the universe is infinite relation. Each monad mirrors and refers to an infinite number of other monads and there is basically no limit in these mirrors. Insofar as each monad retains inner unity in these reference structures, its mirroring is one coherent space-time. Causality and spatio-temporal sequence are thus nothing other than the mirror of monadic unity. Leibniz's proposition that nature does not make leaps is also derived from these considerations: for any leap would literally blow up this unity structure of the monad. The monad would become schizophrenic.

For Leibniz, then, there is a unity of space-time that mirrors the unity of the monad. The continuity of evolution would thus be the biological representation of this inner unity of each monad. It should also be noted that evolution only exists because there are subjects (and vice versa). The highest degree of subject-likeness is reached when unity is attained *as* unity, when self-understanding as a subject is possible. Time is not a large container, but the mirror of the monads, where time and consciousness are inseparable. A particular aspect of Leibniz's conception of time is expressed in his dictum of the "prestabilised harmony of nature and grace". In order to preserve the unity of action, every action must be sufficiently founded. Leibniz's unique perspective, however, is that human action is sufficiently founded in the infinity of God. Man acts freely, as it were, as God's self-fulfilment. Empirically, it can never be determined whether an action is sufficiently founded or not, because to do so we would have to see the totality of all reasons. Leibniz, however, believes that this must be transcendentally presupposed, since otherwise the idea of unity would fall away and the universe would disintegrate into completely incoherent fragments. If we consider ourselves as subjects, which our ethical and noetic experience compels us to do, then we can only be so under the postulate of a sufficiently founded world. Such a world, however, does not experience this justification in empirical time, but in the "infinite", in God, in a future which can never be directly represented ("grace"). Thus, Leibniz places the empirical world ("nature") next to the eschatological, i.e. the world of freedom ("grace"), which transcends chronological time. This happens, where man understands himself as a free Being and also acts accordingly.

4 Hegel: Language, Difference and Time

Hegel elaborates on this concept. Like Leibniz, Hegel thinks of neither the world nor the subject as object-like. World and time are intersubjective, i.e. linguistically and spiritually mediated. They are embedded in the relations between subjects. Therefore, there is no subject-free time, rather time is an expression of these relations. The "I" is not an object, because any objectivity would literally come "too late", since thinking and thought "I" never coincide. Whenever the "I" wants to find itself in the world, i.e. to mirror itself, it experiences itself as withdrawn and time as the difference between the thinking and the thought "I". This difference reveals itself in language, which never directly represents the object, but rather the difference that emerges between the subject and its object. Thus, the object is not directly given, but linguistically mediated. Language, as Hegel shows, sublates the object, i.e. it preserves it, suspends it as an object and raises it to a higher level by spiritualising it. The object is therefore never purely objectively present, but is integrated into the temporal structure of language, which results from intersubjective relations.

Language constitutes the foundation of Being. Since the linguistic expression never completely coincides with its object (when I name the object, this naming does not directly coincide with the object, the word "table" is not the table itself), but is its negativity (*Aufgehobenheit*), every meaning reveals itself in the referential context of language and not directly on the object itself. Leibniz's essential component is the monad, which expresses an infinite relational event that manifests itself temporally and spatially. For Hegel, the essence is the openness of the meaning that underlies every linguistic expression. For him, the subject can never be represented positively, but is that which eludes in every linguistic event. It is the difference (the negativity) of the encountering, called *Concept* (*Begriff*) by Hegel, which can never be fully represented (positivised) but always refers to other differences. Our world, then, is not an assemblage of atoms, but a network of the open, constant differentiation of the *Begriff*, which is not the concept of something positive, but denotes radical openness itself. This openness *is* the subject, and therefore Hegel's world (as well as Leibniz's) is conceived of as subject-like and not object-like.

Time is not an external reference between objects, but the event of the Concept differentiating itself. For Hegel, the Concept and the subject are not to be understood nominally (object-like), but verbally (subject-like), and time is therefore the verbalisation of the Concept, i.e. the subject in its verbal reality, wherein openness (future) and closedness (past) unite in the present to form an open forthcoming past. In our language, this can best be represented by the term future anterior, insofar as this does not testify to an arbitrary event of the future, but to the openness of what will have occurred. “There will have been openness and the freedom associated with it” would be the summary of the Hegelian concept of time.

5 The Alterity of the Eighth Day

Hegel conceives the Being as subject-like. In the network of differences that characterise Being, the difference that refers to itself (i.e. self-consciousness) is the decisive factor. Insofar self-consciousness, as self-reference, is negative *identity* with itself – negative because there is no determinable starting point except for the act of differentiating itself – it is also self-*difference*. Self-consciousness therefore implies radical alterity and is a “becoming-other” in its own self-reference. Ultimately, this alterity is experienced as the withdrawal of any definite self-location in Being, as its absolute negativity. Every Concept, every monad, with Hegel as with Leibniz, is self-withdrawal in self-reference and thus mortal, insofar as *mortality constitutes the most radical form of withdrawal.* The innermost moment of subject-like time or temporal subjecthood is its openness to its other, which reveals itself as withdrawal into mortality. Only the subject is mortal.

The subject can experience this mortality as annihilating death and chaos if it wants to bring time – as this alterity – under control, that is, to deny its own mortality as surrender to its Other in favour of absolute self-control. In this case, it will regard time as something to be overcome. Self-consciousness attempts to “erase” time through the thought of a static eternity or through the denial of its own subjectivity by regarding the world as an eternally chronologically ongoing machine without purpose or meaning. When the subject is objectified, mortality also takes on a different meaning: it is no longer the withdrawal and openness of the subject, but its destruction and transition into nothingness.

The biblical creation story suggests another way of dealing with the withdrawal of time: The seventh, or rather the eighth, day inscribes an openness in time that eludes all representation and functionalisation. This new creation of time, the eschaton, is the transition into the radical openness of time, in which the “I” and its freedom are to be located. The third level of time (time III), i.e. the eschaton, inscribes a moment of alterity, withdrawal, openness into time. Becoming a subject in the biblical sense implies recognising this openness and thus *admitting one’s own mortality, vulnerability, and withdrawal, which is also the withdrawal, vulnerability and mortality of the other. What is celebrated, then, and what gives time a subject-like character, is the moment of otherwise unavailable openness.*

The strange aspect of this "third" time, which is expressed in the resurrection on the "third day" and in the seventh and eighth day, is that it "doubles", or dramatically repeats, death (the night of the "second" time), because, like death, the eschatological time can also only be experienced negatively as Other. If death is present in the mechanical separatedness as entropy and in the life-process as its cyclical becoming and perishing, it shows itself in self-consciousness as (self-)withdrawal. Mankind strives to control this withdrawal. We find the oldest attempt at this in genealogy (Heinrich 1966), wherein man seeks to determine his own past and future, and thus "(re)fill" the withdrawal of time, i.e. live on forever. However, genealogy is strongly challenged when man learns that families and peoples are also subject to transience. Two attempts are therefore found as the ultimate overcoming of time: the first is a metaphysical attempt to decipher temporal existence as "nothingness" and to oppose it with an eternity that is as unattainable as it is untouchable. The second attempt is the inheritance of the first: the static eternity becomes the static time quantum by quantising time mechanically and thereby bringing it to a standstill and making it calculable. God is replaced by the machine, which has its deepest form of expression in perpetuum mobile as an infinite causal chain. The world becomes an eternally linear, potentially calculable function that knows no end. Consequently, this understanding of time leads to nothingness, which erases the intrinsic value of all structures. The consequence of this is visible in today's cosmological simulations, as well as in the way we let our dead be symbolically replaced by the horror of an absolute void.

The biblical creation story instructs us against bringing the withdrawal of time, and thus death, under our dominion. The decisive turnaround occurs in the epiphany of the time of the Other, which is connected with the displacement from one's own will to control. The feast "responds" to this epiphany of time offering the joy of self-donation. Consequently, one's own mortality, which is completed in the encounter with the Other, is celebrated – in this sense, any celebration that seeks to displace one's own mortality by means of immediate enjoyment would be a perversion of the feast – and one's own existence transcends into an event that is more original than one's own self-positing. The mortality of the subject does not imply its extinction, but the end of the possibility of objectifying the subject. This opens up the subject as an Other and A-Present that transcends Being towards a past and a future that precedes every present. This pre-gift of time is expressed by the eighth day, whose future is the beginning of time.

References

Bahr, Hans-Dieter. 2008. *Zeit der Muße – Zeit der Musen*. Tübingen: Attempto.

Heinrich, Klaus. 1966. *Parmenides und Jona*. Frankfurt am Main: Suhrkamp.

Klein, Hans-Dieter. 2006. *System der Philosophie*, Vol. 2: *Naturphilosophie*. Frankfurt am Main: Peter Lang.

Koch, Anton Friedrich. 2006. *Versuch über Wahrheit und Zeit*. Paderborn: Brill.

Strickland, Lloyd. 2014. *Leibniz's Monadology. A New Translation and Guide*. Edinburgh: Edinburgh University Press.

Zenger, Erich. 1983. *Gottes Bogen in den Wolken. Untersuchungen zu Komposition und Theologie der priesterschriftlichen Urgeschichte*. Stuttgart: Katholisches Bibelwerk.

Andrea Toniolo

Time, Revelation or Negation of the Eternal? The Modern Metaphor of the "Death of God"

Abstract: Before the modern physical 'revolution' on the conception of time/space, it was the modern theological (and philosophical) thought that threw the 'classical' view of time and history into crisis. This crisis is emblematically expressed by the metaphor of 'the death of God' (Nietzsche, Hegel and Jüngel), which challenged the classical ontological view of being and of God. Such a metaphor made it possible to think of time and history as a revelatory place of God's eternity.

> God's eternal being is more diverse and even more temporal than we are capable of thinking.
> (Jüngel, *Possibility of God in the Reality of the World*)

The "passing of time" (the quantitative conception of time, synonymous with finiteness and transience) has been interpreted – in the Western philosophical and theological tradition, marked by Greek thought – as a negation of eternity, as diametrically opposed to it, or at best as a "fertilizer" for eternity. According to this perspective, the meaning of human existence, subjected to transient time, emerges insufficiently; it is denied any form of transcendence and death is considered man's abyss, the confirmation of his meaningless finitude.

The classical conception of physical-mechanical time (as opposed to "eternity"), has conditioned man's understanding and led to thinking of temporality ("being in time") as a limit to be overcome. This conception of time, lacking consistency and permanence, and therefore non-being, also conditioned the very idea of revelation, i.e. the way in which God enters into relationship with history. Scholastic theology had thought of two forms of revelation: natural and supernatural. The first is that of reason and consists in recognizing the signs of the creator in creation (for the simple fact that creation alone cannot be explained). The second is that which comes directly from God through biblical history: God reveals Himself by speaking personally to man. The two parallel ways were the result of a different and opposed conception of natural reality (temporal, finite, limited, non-existent) and supernatural reality (eternal, infinite, immutable).

The modern rationality sees in the supernatural path a violence of reason, an imposition to be freed from as irrational, and therefore unacceptable. Reason and

https://doi.org/10.1515/9783111313610-015

revelation become incompatible. The parable of modernity is perpetually marked by this conflict, which Hegel interprets with the image of the "unhappy conscience". Thus, before the "revolution" of modern physics on the conception of time/space, it was modern philosophical and theological thought that threw the "classical" view of time and history into crisis. The modern crisis of the philosophical and theological tradition is emblematically expressed by the "death of God".

1 The "Death of God" as a Metaphor for the Aporia of Thinking about God in the Modern Age

The announcement "God is dead" immediately brings to mind Friedrich Nietzsche's well-known page in *The Gay Science*: in just a few passages, the German philosopher represents all the modern drama of that expression. It indicates the end of a world of representations. But I do not stop to comment on that text. I prefer to accord more relevance to the Hegelian reading of the death of God, taken up by some theologians, and I show its relevance to our theme.

Hegel – the young Hegel – coined the expression "speculative Good Friday" to indicate that in the feeling of the lost God, proper to modernity, the historical Good Friday has been elevated to a concept, has been understood (*begriffen*): the modern subjectivity (the exaltation of the finite self) experiences the same absence of God as Jesus of Nazareth when he died on the cross. But it is from this feeling and suffering that the yearning and desire for God is reborn, according to Hegel. The death of God becomes the interpretive key not only to modernity, but also to the whole of reality, which is a dialectic of death and resurrection. The death of God represents the split reality of history and the finite spirit (the unhappy consciousness), and at the same time the rational power of the absolute Spirit (God) to take on the abyss of negativity. It expresses the power of reconciliation (*Versöhnung*) of the other: "It is the consciousness of the loss of all *essentiality* in this *certainty* of self, and of the loss of precisely this knowledge of self, – of substance, as well as of the Self; it is the pain that is expressed in the harsh word: *God is dead*" (Hegel 1973, 255, my translation). In these passages the three meanings of the "death of God" according to Hegel emerge: as "infinite pain" because of the loss of God, as "negativity in God", as "absolute reconciliation".

According to Hegel, time (the history with all its negativity, finitude, evil, pain) is the revealing place of God, of his essence. In negativity, the consciousness of the Spirit emerges: "God corresponds to the infinite consciousness in the moments when it learns to understand its contingency as an opening to the Absolute" (Appel 2018, 145; my translation).

Allow me, here, a brief but illuminating reference to Schelling, according to whom only "the ecstasy of reason" (Appel 2018, 140), i.e. the abandonment of autonomous reason, represents the way out of temporality as limitation (and death is its greatest sign): "this ecstasy is possible when God reveals himself as the eschatological beginning of time and for this reason God's salvific will is not philosophically deducible, but only witnessable in the *kenosis* as the response of a preliminary historical self-revelation" (Appel 2018, 140–141; my translation).

It was above all the evangelical theologian Eberhard Jüngel who would show the philosophical and theological implications of the metaphor of the "death of God" (metaphor in the etymological sense, i.e. as a transfer of meaning: the death of God operates a transfer in the way of thinking about God), going beyond Nietzsche's more widespread but limiting meaning, according to which the death of the metaphysical God constitutes the overcoming of a metaphysical concept of God that has become unbearable for modernity, since it is incapable of reconciling eternity and time, immutability and mutability. In short, the announcement that there is no longer any God.

On the other hand, a properly theological discourse of the death of God, takes us back to the thought of Hegel, who, as highlighted above, makes the death of God the center of a metaphysical discourse of God. He speaks of the negation not of God, but in God himself, within whom the inversion, the death of death, takes place. God's death is the revelation of God's essence, which is infinite love, that is, the ability to identify with what is foreign to him in order to reconcile him: "Hegel reminds us that the word of the cross speaks of God himself, that the event of Jesus Christ's death invests God's being in an absolute way" (Jüngel 2005, 38; my translation). The Tübingen theologian shows the true Christological (not just modern cultural, identified with atheism) origin of God's death. I must say, however, that it was modernity that resonated or resurrected it.

Jüngel has the merit of Christologically anchoring the metaphor of God's death, showing two implications: its significance for God's being and for man's being. It becomes a revelation of God and man. I take them up with the perspective that marks our conference: time and eternity.

1.1 God's Death for God

The biblical texts attest unambiguously that God's death on the cross is not one event among others, a passing event in view of the resurrection, but the event that determines God's very being (it cannot simply be thought of as the eternal opposed to time, *spiritus infinitus* opposed to the finite). God essentially revealed Himself as God precisely through the death, which represents the expression of finiteness and transience, of limitation and temporality, of strangeness in relation to the infinite. The death was not a price to be paid, a temporary passage in view of something else, like "through an arch of triumph" (Jüngel 2005, 47; my translation). God defined himself through that death, that suffering, and removed from it the character of strangeness, of distance, of rupture:

> The essential action [extraneousness] removed from the death determines the living being of God as granting an eternal place to those who, existing *en Xristò*, are elected and destined for the eternal being of God. God's being, therefore, is no longer to be thought of as *omnino simplex esse*. God's eternal being is more diverse and also more temporal than we are capable of thinking (Jüngel 2005, 48; my translation).

This is because the death has become "a predicate that determines God's being. God suffers the death in order to show himself as God precisely in this act of suffering" (Jüngel 2005, 48; my translation).

The "death of God" as an image that evokes the fate of Western metaphysical thought, rooted on the opposition between time and the eternal, does not mean the end of metaphysics (the death of the metaphysical God, as Nietzsche cried out), but, following Hegel, it means thinking the being of God (and being as such) from the death of God: "the last thought of metaphysics is not the concept of the death of the metaphysical God, but the metaphysical concept of the death of God" (Jüngel 2005, 50; my translation).

1.2 The Death of God for the Human Being

Every theological discourse has anthropological relevance (and we agree with Feuerbach), it implies an understanding of mankind and the world, which also has repercussions in the fields of science and philosophy (distinct but not separate). What is the meaning of God's death for the human being's death? The death insofar as it has become "of God" cannot be left "behind", one cannot think of life without death. Death, by virtue of God's death, is recomposed, redetermined, because it remains the event of life: "God, once in death, no longer disposes of death, because death can no longer dispose of God" (death is deprived of its nonsense, of its being an end). It is not left behind but brought into life itself: "for this reason, death is transformed into a blessing, which makes it possible to be able to die" (Jüngel 2005, 51; my translation), it is accepted in life. It is no longer understood as the absence of relationship (rupture) but as relationship, since in death life is manifested (the possibility of revealing an authentic, donated life), relationship is maintained (no one dies any longer for himself, but with and for).

Death is no longer alienating, extraneous but proper to man as an act of life. It is certainly a "limit" of the human being, but a limit that qualifies the human being as such, a limit that is not negative, but a condition of relationship. On the contrary, it becomes a boundary (*Grenze*), which limits but also relates. Death is therefore, "as the boundary of life, the end of a time of life and a space of life, but not the end of time and space" (Jüngel 2005, 183; my translation), which in eternity are an indistinguishable unity.

In an essay titled "Boundaries of the Human Being" ("Grenzen des Menschseins"), written in the form of a thesis, Jüngel repeatedly dwells on the relationship between time and eternity (or rather time, space and eternity). The human being, as a difference to God and others, is "limited by eternity" (Jüngel 2005, 178; my translation), which is "the necessary divine form of the relationship between God and himself and between God and the human being" (Jüngel 2005, 178; my translation). The fact that the human being is constituted by the relationship is the eternal limitation; without this limitation man would no longer be human: "man, insofar as he is eternally limited by the grace of God, is eternal" (Jüngel 2005, 179; my translation).

Wanting to be without limits means wanting to be without relationships (*homo peccator*). This eternal limitation has a "worldly" configuration (being-in-the-world), given by time and space, which are the condition of its existence, and thus of its history, its progress. The eternal limitation is expressed in the worldly (through time and space), "so that Gods love for the human being can happen as God's history with the human being" (Jüngel 2005, 180; my translation). This worldly history is the "silent parable of eternity" (Jüngel 2005, 181; my tranaslation), where the inseparable and indistinguishable unity between time and space takes place.

The gain of contemporary philosophy and theology is to have considered temporality as a unique and original qualification of the human being (*homo temporalis*, Jean-Luc Marion would say), not diametrically opposed to eternity, but correlated with it (man is in his creatureliness *imago Dei*). The time that marks human history (and of the cosmos itself) is not canceled or annulled, but "welcomed":

> The new mode of existence that the human being receives as a gift from God has fully and totally to do with time: it is constituted by time, precisely because everything that has been experienced as present, in the laceration of earthly time, is carried and welcomed in it. The existence transfigured before God is "received time" (Lohfink 2020, 206; my translation).

2 Time and Eternity Include Each Other

In contemporary theology, thanks to the notion of revelation as God's "self-communication" in the history (more anchored in the biblical conception of time), temporality is not understood as other than eternity, but as the irruption of eternity, an expression of it, as the form in which the Triune God reveals himself. God assumes temporality as a manifestation of his own eternity: "God's self-revelation in Christ brings about a communion between God's eternity and the temporality of creation" (Martin 2009, 102). This means that the temporality is contained in the very being of God the Trinity.

The conception of time in Barth – which has just been mentioned – does not only have theological or philosophical references but may also have similarities with the conception of time in modern physics, which relativizes the absolute magnitude of clock time:

> Although Barth's reference was from the standpoint of theology, his statement that "there is no such thing as absolute time, no immutable law of time" is true within a contemporary understanding of time as part of the physical creation. The enigma of time is not an issue confined to theology and philosophy and, in modern physics, time and eternity are now understood within a framework far removed from the ordinary linear perception of past, present and future. Barth's statement above is upheld by the theory of Special Relativity in which there is no universal "present" nor an absolute uniform rate for the passage of time (Martin 2009, 107–108).

Thus, the affinity between Barthian theology on time and the theory of relativity is argued. At the same time, the distance from the purely "subjective" conception of time is emphasized (cf. Augustine's memory, Kant's a priori form, Merleau-Ponty's phenomenological approach), according to which it would have nothing to do with "objective" time (according to physics, it is only the succession of individual instants).

In the biblical world, "objective" time, i.e. historical-cosmic time ("when the fullness of time came", Galatians 4:4; "The creation waits eagerly", Romans 8:19) is decisive for the revelation of God and for man's understanding. There is a close connection between "subjective" time and historical-cosmic time:

> An awareness of linear time is, for human beings, an essential part of their nature: it is "no mere idiosyncrasy of human psychology but an intrinsic feature of reality . . . " [quotation of J. Polkinghorne] It has been pointed out that the God of the Old and New Testaments demonstrates "a deep engagement with the historical process, with the becomingness of the world" (Martin 2009, 108).

According to Swiss theologian Hans Urs von Balthasar, by virtue of the incarnation and resurrection "the eternity is 'simultaneous' with every moment of man's temporal being" (Balthasar 2017, 63; my translation). There is a "now of eternity. [. . .] eternal life already begins in the now" (Balthasar 2017, 64; my translation). In Christianity, therefore, existence in time will not be a memory of the past, something that remains behind, the ephemeral that has no value. Existence in time is already the figure of eternity, because redemption is of time not from time, of the body not from the body. Eternal is this time, this world, this body.

Time and eternity include each other: eternity is not the future but the present time, it is the eternalization of time. In John's gospel, the *eschaton* (eternal life) is not the future but the present: "He who hears my word and believes in him who sent me has eternal life and does not go to judgement, but has passed from death to life" (John 5: 24). The fourth gospel transfers the future into the present: he who believes already has eternal life. What is called "eternal life" is not something else and totally new, but the fulfillment of the present time, of the hereafter: "what has not happened in this history cannot be resurrected, cannot be brought into eternity with God" (Lohfink 2020, 253; my translation). Even "what was lacking in love here, cannot be magically conjured up there. [. . .] What is built in time, is built forever; what is omitted, remains omitted" (Lohfink 2020, 253; my translation). The eternity, according to Christianity, begins here and now.

In Balthasar, there is a profound analogy (correlation) between time and eternity, or rather between time and the Trinity: the dynamics of time (becoming, suffering, giving) are already inscribed in the very life of God. In contrast, "the temporal life, in order to truly show a nexus of meaning and not unravel into clear incomprehensibility and incoherence, absolutely and in all cases needs a summarizing attribution of meaning, which can only result from a state of eternity" (Balthasar 2017, 75; my translation). Kant had recourse to eternal categories to ground morality. But existence itself, in order not to fall into absurdity, needs an eternal foundation: it is the sense of existence, of being called into cause.

The temporality, therefore, qualifies the human essence for two reasons: as "memory-time", i.e. time as the memory that constitutes man's consciousness (cf. Augustine, Ricoeur: identity is the fruit of memory, which is self-consciousness in time); and, secondly, as relationality, i.e. man becomes himself in the relationship with others, implemented in historical time. Temporality is thus the place where freedom and responsibility are exercised.

This is why every moment of time that marks existence is decisive for the realization of the human being (memory and relationship) in its identity and thus for eternity, which is already realized in the present hour (cf. John 10:28: "I give [present time] to them eternal life"). "Every moment of our life takes on an extraordinary weight. For what is not experienced here, will not be there in the hereafter. What is lacking in love here, will not be magically evoked there" (Lohfink 2020, 253; my translation).

What is built here and now, within time, remains forever. It is eternal.

References

Appel, Kurt. 2018. *Tempo e Dio. Aperture contemporanee a partire da Hegel e Schelling*. Brescia: Editrice Queriniana.

Balthasar, Hans Urs von. 2017. *Escatologia nel nostro tempo. Le cose ultime dell'uomo e il cristianesimo*. Brescia: Editrice Queriniana.

Hegel, Georg Wilhelm Friedrich. 1973. *Fenomenologia dello spirito*, Vol. 2. Scandicci: La Nuova Italia Editrice.

Jüngel, Eberhard. 2005. *Possibilità di Dio nella realtà del mondo. Saggi teologici*. Torino: Claudiana.

Lohfink, Gerhard. 2020. *Alla fine il nulla? Sulla risurrezione e sulla vita eterna*. Brescia: Editrice Queriniana.

Martin, Hilary C. 2009. "Eternity and Temporality in the Theology of Karl Barth". *Science & Christian Belief* 21: 101–110. https://www.cis.org.uk/serve.php?filename=scb-21-2-martin.pdf, last accessed 21 September 2023.

Piero Benvenuti

Cosmology and Cosmologhia: A Much Needed Distinction

Abstract: In discussing the concept of eternity from the perspective of our current knowledge of the reality, it is important to draw a clear distinction between Cosmology, the rational model of the Cosmos as it can be constructed by the correct use of the scientific method, and *Cosmologia* by which I mean a global comprehension of the whole reality, with no distinction between matter and spirit, between what can be modeled and understood by the scientific method and what needs additional forms of epistemology. Only within this expanded *Cosmologia* it makes sense to argue about the concept of eternity.

Years ago, I was invited to give a talk about the nature of the universe to the students of a high school in a capital city of southern Italy. When the students were entering the room, I noticed that one of them was a Down person: I have a special feeling for the Down people, therefore I gave him a welcome wave while he was proceeding to seat in one of the back rows. As usual, after giving my talk, I had to answer several questions posed by the students, most of them about black holes, dark matter, dark energy . . . It seems that the dark side of the universe is the most intriguing one . . . When I was about to wrap up the conference, I saw my Down friend raising his hand. In the characteristic lilt of those gifted people he said, almost verbatim: "Professor, you showed us how the universe is composed with its stars and galaxies and you explained us its evolution, can you now tell us why it exists?"

This anecdote, which is always vividly present in my memory, serves as a perfect introduction to the subject of my paper: the much needed distinction between Cosmology, intended as the rational model of the Cosmos as it can be constructed by the correct use of the scientific method, and *Cosmologhia*[1] (I am reading the Latin word with the classical pronunciation in order to keep the distinction clear also in Italian). With this term I intend a global comprehension of the whole reality, with no distinction between matter and spirit, between what

1 *Cosmologhia*, as a distinct term from Cosmology, has been proposed by the cosmologist George F.R. Ellis (Ellis 2017).

Article note: This paper was presented to the conference "Eternity between Space and Time: From Consciousness to the Cosmos", Padua, 19–21 May 2022.

https://doi.org/10.1515/9783111313610-016

can be modeled and understood by the scientific method and what needs additional forms of hermeneutics: philosophy, theology, poetry and art.

As I will try to demonstrate in this paper, this distinction is particularly relevant for the theme that has been proposed for this relevant session of the Conference: eternity and physics (here, hoping I am not disappointing the organizers, by physics I mean more generally "physis" or nature).

Let me begin with a short consideration on the ancestral origin of the concept of eternity. Since the early emergence of the human consciousness, men and women were confronted with the obvious finiteness of their earthly life to be compared with the existence of a Cosmos which was unattainable and apparently lasting forever. Indeed not only the most common cosmic phenomena, like the day-night diurnal cycle or the monthly lunar phases, seemed to repeat themselves with a relentless regularity that was reported across generations and generations, but especially the more subtle cyclic phenomena, like the 18-year Saros cycle of the eclipses or the precession of the constellations that required centuries of careful observations to be revealed, created the solid belief that the cosmos and its movements were eternal.

It is reasonable to think that, beyond other possible considerations, this comparison induced the human aspiration to exceed the limit imposed by the physical death, hoping that consciousness, or soul, could continue its existence in a different dimension if not coincident with the timeless cosmos.

This somehow undefined human longing, which is to be found in all ancient civilizations, obtained a more solid foundation in the Aristotelian cosmology, which made a clear separation between the mutable sub-lunar world and the imperishable empyrean, the seat of the eternal and perfect – either rectilinear or circular – cosmic motions. It is important to remark here, for the later discussion on space and time, that for Aristotle time and space are not absolute and self-existing, rather they are linked to motion. Without motion time would not exist, therefore the empyrean, or the sky, to use a more popular name, became the obvious synonym of eternity also in view of its substantial or quintessential distinction from the perishable earthly world.

It is to no surprise that the scholastic theology, embracing and reinterpreting the Aristotelian philosophy and cosmology, almost physically identified the Kingdom of heaven, the Paradise, with the empyrean, where the resurrected would eternally enjoy the afterlife.

We know that the crystalline spheres of the Aristotelian cosmology were definitely destroyed during the fatal nights of December 1609, with the first observations, not far from where we are now, by Galileo Galilei with his famous "cannocchiale". The empyrean and its quintessence were demoted to normal earthly matter and the cosmic motions, albeit long standing, lost their essential eternal characteristics.

While for Astronomy a new era had begun, Cosmology entered into a deep crisis that was to last for almost four centuries: the cosmos could still be thought as eternal, but its eventual eternity became totally disconnected from the human consciousness. The new cosmic agony is well represented by the verses by John Milton in his most famous poem, *Paradise Lost*, which he composed after having visited in Arcetri an elderly Galileo. Referring to the starry sky, which has become a stranger to Man, he wrote:

> How Nature, wise and frugal, could commit such disproportions, with superfluous hand so many nobler bodies to crate? (*Paradise Lost* 8, 26–28)

Humanity was lost again in a universe that has become disconnected in form and purpose from it.

While the very foundations of the Aristotelian cosmology were so clearly collapsed, the theological and hence the popular concept of a designed physical location for the eternal afterlife remained almost unaltered, without realizing it had become suspended over an undefined vacuum. It should be remarked here that the core of the Galilean/Newtonian revolution was the introduction of the new concepts of absolute space and time which were essential elements of the new theory of motion and well as the basis for the universal gravitation theory. It is a pity that the philosophical and theological implications of the new definition, which were the subject of the vivid exchange of letters between Samuel Clarke and Gottfried Wilhelm Leibniz, were soon forgotten. The evidence that the medieval theological vision of the afterlife remained unaltered in spite of the 17th-century revolution, can be found in the formulation, in 1950, i.e. only 72 years ago, of the catholic dogma of the Assumption of the Virgin Mary. The official text, which makes several references to the scholastic theology, concludes: "we pronounce, declare, and define it to be a divinely revealed dogma: that the Immaculate Mother of God, the ever Virgin Mary, having completed the course of her earthly life, was assumed body and soul into heavenly glory" (Pope Pius XII, 1954), assuming therefore that there is somewhere in the cosmos a place where the resurrected bodies and souls will enjoy the eternal afterlife (multiverses were still to come!).

We can concede that, at the time of the proclamation, the new cosmology was in its infancy, but today the dogmatic theologians should seriously consider reformulating the text in such a way it becomes understandable, without losing its profound theological content.

I presented you this very concise historical review in order to highlight the intimate relation between the concept of eternity, either referred to space or time, and our perception and understanding of the cosmos.

As we have seen, after the Galilean revolution, humanity was left bereft of an understandable cosmology for almost four centuries. It was only 95 years ago, with the publication of the seminal paper by George Lemaître, that the new cosmology began to take shape. Since the very beginning the new model of the cosmos revealed its most fundamental and unexpected characteristics: the universe is essentially evolutive, it has a history that saw it going through phases quite different from each other, but all intimately interconnected by a unitary process that produced and is producing entities and phenomena of increasing complexity.

During the last decades, the new sophisticated observing facilities, operating on the ground as well as from space, allowed the cosmologists to reconstruct the cosmic history with an astonishing precision along a path of about 13,8 billion years. It is not my intention here to describe in detail the current cosmological model, I will limit my consideration to its most innovative and essential feature: its space-time evolution. Indeed, this is the characteristics that really matters when discussing today the concept of eternity.

However, before entering into the core of our analysis, a few preliminary remarks are in order, all descending from the cosmic epistemology.

The first remark concerns the new revolution about the space-time and cosmic time that took place in rapid sequence first in 1905 with the special relativity theory by Einstein and later in 1927, with the above-mentioned solution of the general relativity equations applied by Lemaître (and independently by Friedman) to the entire universe. The special relativity forced us to abandon the Newtonian concepts of absolute space and time: different cosmic experimenters, when communicating to each other the results of similar experiments have to take into consideration the mutual velocity of their local laboratories in order to make sense of the comparison. We had to surrender to the reality of dimensions and time-laps that change if observed from different reference systems. It would be indeed problematic to compare eternity with a relative time. However, a second counter-revolution came with the discovery of the cosmic time, which can be thought as the clock-reading that can be assigned to each evolving phase of the expanding universe. Paradoxically, this cosmic time *can* be considered absolute – at least within the universe it describes – in the sense that any cosmic observer can in principle determine the age of the universe he lives in, by measuring, for example, the apparent temperature of the cosmic background (here and now we call it the cosmic *microwave* background, but other cosmic observers, living in a different galaxy and in a different cosmic epoch, may see it peaking at different redshifted wavelength). We should conclude that, when discussing eternity, we should compare it to the flow of the cosmic time, leaving the relativistic space-time aside.

A second and more relevant remark concerns the applicability of the Galilean scientific method in the construction of a global model of the universe. The question might appear paradoxical: it is thanks to the modern scientific theories, in particular to quantum physics and to the general relativity that we have been able to successfully describe a wealth of cosmic phenomena that occurred in the various phases of the cosmic evolution. Some of these phenomena, like the black holes and the gravitational waves, are so intrinsically complex and their physical conditions so far away from what we can hope to experiment in our terrestrial laboratories, that we can only look back in wonder about the effectiveness of the scientific method in modeling the reality. Why should we then raise doubts about its applicability to the entire cosmos?

The reason is very simple: because the cosmos, seen as a global and unitary phenomenon, is unique: we cannot rewind the cosmic history and start it again modifying its initial conditions. That is what normally physicists do when they perform experiments: the repeat it by changing the initial conditions in such a way that some regular patterns would emerge to be called later physical laws. The repeatability of the experiment is an essential feature of the scientific method because it allows the physicist to separate the effect of the initial conditions from the underlying physical laws. Obviously this classical procedure cannot be applied to the cosmos as a global unitary experiment: not only we cannot modify the initial conditions of the emerging universe, but we cannot even determine what they were.

One may object that the same situation applies to the study of all celestial phenomena, like the Sun, a planet or even a black hole. However, the inability of modifying the initial conditions is amply compensated by the possibility of observing a large number of objects belonging to a specific class, like, for example, the stars. By observing millions of stars with different masses, dimensions, temperature, chemical composition, etc. astrophysicists were able to learn a lot about the structure of the stars, their formation and evolution.

Unfortunately we do not have the possibility of investigating the properties of a cosmos other than the one we live in (I will talk about the multiverses hypothesis in a moment). Consequently, we cannot effectively distinguish between the initial conditions and the physical laws that were guiding the initial phases of the cosmic evolution. This epistemic crisis is exacerbated by the difficulties encountered by the theoretical physicists in combining in a single compatible theory the two main pillars of our rational understanding of reality: quantum physics and general relativity. And this morning we learned a lot about this strenuous fundamental effort.

While we must admire these important attempts to dig deeper and deeper in our ability to understand reality, we have to admit that we have an elephant in the room: given the above considerations about the unicity of the cosmos, shouldn't we recognize, with Galilean humbleness, that the scientific method alone is inadequate to investigate the origin and evolution of the universe? Or, in this fundamental quest, shouldn't the scientific method delivers its findings to philosophy and theology to complete the task? In other words, shouldn't the cosmologists become, or return to be as in the old times, philosophers of nature?

Waiting for this epistemological revolution to happen, cosmologists have reacted to the crisis by proposing models that circumvent the issue of the initial conditions as well as of the beginning. Our universe could be just one of the many or infinite possible universes, each of them with diverse initial conditions and possibly physical laws and evolution histories. Or else, our universe could have a cyclical history, avoiding therefore to have a beginning. These are fascinating hypothesis that in fact gained high popularity among the experts as well as the layman: they are indeed very interesting and scientifically plausible, but they all lack intrinsically the possibility of being verified by observational data. Parallel universes cannot send messages across their respective boundaries nor a cyclic universe can send us news about its previous existence.

These hypothesis, which could be considered attempts to exit from the current cosmic impasse, are not scientifically verifiable, or, more technically, falsifiable, therefore they belong to the wider category of philosophical or theological proposals. As such, they require a free and personal act of faith to be accepted.

If this is the situation, we can now return to the core of my intervention which is highlighted by its title: cosmology versus *cosmologhia*.

Considering the incontrovertible discovery that the cosmos has a history and its evolution is responsible for the emergence to existence of entities and phenomena of ever increasing complexity, most of them unpredictable a priori, and that the most complex and intriguing reality that emerged at the tip of a 13,8 billion years journey is the human consciousness, isn't logical and convenient to move up from the pure scientific cosmology (or better said, evolutionary astrophysics) and embrace the concept of a global cosmology, or *cosmologhia*, that, as in the old classical times, includes human consciousness and its cosmic destiny in its considerations?

If we dare to take this bold step, the relation between *cosmologhia* and eternity becomes much more interesting.

In fact, if we would limit our discourse to the pure scientific cosmology, we could indeed think about the cosmic eternity as the limitless temporal evolution of the cosmos, that, from what we learned so far, does not seem to have an end. However, that approach has little, if any, anthropological interest: we know, scientifically, that humanity, similarly to our individual terrestrial life, will face an ineluctable end when the Sun will become a red giant and will swallow all the planets (by the way, after having roasted them, starting from the closer ones!). It is an end very, very far away in time, but still an unavoidable one. Our aspiration to eternity and the eternal cosmic evolution of our universe (or multiverse or cyclic universe) do not have any relation between them.

As already mentioned, the discourse becomes much more interesting if we take into consideration the possible global cosmological models. These various *Cosmologhiae* will all have as a common, solid and necessary pedestal in the findings of the (verifiable and falsifiable) scientific models of the cosmic evolution. But they will differ by the choice of solution of the stumbling blocks that, as we have seen, the scientific method encounters. These can be represented by the multiverse hypothesis, the cyclical universe, or other philosophical or theological hypothesis: in this approach all of these cosmological models share the same scientific credibility (because of the initial common assumption) and they have all the same trustworthy dignity: however, the motivation to choose one of them cannot be the result of a demonstration, like we do with a geometrical theorem, rather it requires a free personal act of faith.

I may, for example, believe that the cosmos and its evolution, including the emergence of life and of the consciousness, is simply due to chance: I am leaving by chance in the singular universe, among the infinite possible ones, which is compatible with life. Taking this choice, as we have seen, cosmic eternity and eschatology are sitting on incommunicable layers.

Alternatively, and I wish to emphasize this, with equal dignity, I may choose a model in which the entire cosmos and its progressive evolution towards the consciousness is the result of a free act of love which keeps in existence the entire reality, matter and spirit, and is waiting patiently that the emergent consciousness, with an equal free act, recognizes it and reciprocate the unconditional love towards all humanity and the entire cosmos, becoming in this way a partner in the cosmic existence. This profound cosmic relationship, which I would dare to call Trinitarian, lives forever in the Kayrós and will resist to the insults of Krònos, as knows very well whoever has experimented the material loss of some deeply beloved one. Eternity may begin there.

References

Ellis, F.R. 2017. “The Domain of Cosmology and the Testing of Cosmological Theories”. In *The Philosophy of Cosmology*”, edited by Khalil Chamcham, Joseph Silk, John D. Barrow, and Simon Saunders, 3–39. Cambridge: Cambridge University Press.

Milton, J. 2008. “Paradise Lost”, Oxford World Classics, ed. by Stephen Orgel and Jonathan Goldberg. Oxford/UK: Oxford University Press.

Pope Pius XII, “Ineffabilis Deus” https://www.vatican.va/content/pius-ix/it/documents/18541208-costituzione-apostolica-ineffabilis-deus.html, last accessed 18 October 2023.

Gabriele Gionti S.J.

God and the Big Bang: Past and Modern Debates between Science and Theology

Abstract: A short phenomenological account of the genesis and evolution of the universe is presented with an emphasis on its primordial phases as well as its physical composition, i.e., dark matter and dark energy. We discuss Einstein's theory of General Relativity and the birth of modern relativistic astrophysics. We introduce the Big Bang theory of Mons Lemaître and the competing theory of a Steady State Universe of Fred Hoyle. Pope Pius XII delivered a message to the Pontifical Academy of Sciences in 1951 claiming a certain agreement between the creation account in the Book of Genesis and the Big Bang theory (a concordist view). Lemaître always kept separate the scientific and theological planes as two as two complementary *magisteria*. Similar kinds of tensions, between science and theology, emerge also today with the Hartle-Hawking solution to the Wheeler-DeWitt equation in quantum cosmology.

1 Introduction: A Brief History of the Universe's Evolution

The standard model of cosmology predicts that our universe began with a very dense and hot phase of matter and energy that has been called the "Big-Bang" (Stoeger 2010, 155–175), with the density and temperature of this initial phase approaching infinity. With time (between one and three minutes), this hot "plasma" universe began to expand and cool, reaching the temperatures necessary to form the nuclei of the first atoms: hydrogen and helium with traces of deuterium. Evidence of this primordial nucleo-synthesis can be gauged from the fact that helium accounts for nearly 24% of the weight of the baryonic matter in the universe, an amount which could not be produced only in stars. The photons of the early radiation could not escape because they were scattered by the free electrons. When the temperature of the universe cooled down to 4000 K (Stoeger 2010, 155), the hydrogen atoms were able to recombine and the energy radiation of the hot gas was allowed to escape. This primordial radiation was emitted around 380,000 years after the Big Bang, and can be detected even today as the Cosmic Microwave Background Radiation (CMB). This primordial radiation in the universe bears the imprint of its last scattering from the ionized matter. The CMB today is

https://doi.org/10.1515/9783111313610-017

homogeneous and isotropic and has a constant temperature except for very tiny fluctuations. The fact that the CMB has these tiny variations in temperature implies that the ionized gas, from which CMB originated, had very small density fluctuations. These tiny density fluctuations, under the effect of the gravitational field, collapsed and formed the first structures in the universe, galaxies, and clusters of galaxies. The formation of structures happened around one million years after the Big Bang.

As regards the fundamental interactions in the history of the universe, at the beginning of the universe, between 0 and 10^{-43} sec, that is the Planck time at a temperature above 10^{32} K, quantum mechanics and the gravitational field were unified, most probably, in a unique theory called Quantum Gravity. Many scientists believed that quantization of gravity happened by the unification of all fundamental forces, including gravity, into one fundamental interaction. Although this is plausible, it has not been proved yet. It could be, pairwise, that only gravity was quantized in the Planck era. Space and time were quantized as well. The gravitational interaction is the only interaction that exists at Planck scales and very high temperature, all other interactions will disappear. This is a possible scenario as well. Above the Planck era, the gravitational field was present but not quantized, space and time and all other fundamental interactions, unified in one fundamental interaction called GUT (Grand Unified Theory) (Stoeger 2010, 156), emerged. The GUT is a quantized theory. This phase of the universe happened around a temperature below 10^{32} K. The GUT is the quantum unification of the strong, weak, and electromagnetic interactions into one unified interaction. This unification too is far to have been experimentally proven yet, but it looks plausible to many theoretical particle physicists. While the universe expanded and cooled down, we reached a temperature of 10^{26} K. The strong interaction separated from the electroweak interaction. Before reaching this temperature, the universe underwent a period of very rapid expansion, an exponential growth, called cosmological inflation. Cosmological inflation was caused by dark energy, an ingredient of the universe of which we will talk later. Around 10^{15} K the electromagnetic interaction decoupled from weak interaction. The expansion and cooling of the universe from the Planck era to 10^{15} K lasted one minute. From one to three minutes the universe behaved like a gas, as we have already highlighted above.

2 General Relativity and the Birth of Relativistic Cosmology

The universe expands according to the laws of Einstein's General Relativity. This theory of gravitation was elaborated by Albert Einstein. Albert Einstein is rightly considered the scientist who, more than any other, made fundamental contributions to physics in the 20th century. He tried to formulate a theory of gravity that was a field theory and in which there was no action at a distance, as in Newton's theory. For Newton, a massive body "immediately" feels the presence of another body, and it can be said that this system implies a propagation with an infinite speed of gravitational perturbations.

Obviously, Einstein knew Maxwell's theory of electromagnetism well and wondered how gravity could be described, not as an action at a distance, but as a field whose perturbations propagate with the speed of light. It took him 10 years from the discovery of special relativity to answer this question, and finally, in 1915, he came to the formulation of general relativity. This theory represents, in the history of physics, the beginning of a marriage between physical theories and complex mathematical theories. In fact, general relativity would not exist without Riemannian geometry, or rather without Lorentzian geometry (Weinberg 1972, 19). The theory of general relativity is based on one fundamental postulates. This postulate says that the gravitational mass of each body is equal to its inertial mass, that is, the numerical value of the mass by which two bodies attract each other is equal to that of the inertial mass, which indicates how a body opposes movement. A "corollary" of it is the principle of covariance, according to which the laws of physics are the same, therefore covariant, in every reference system. In particular, this theory includes non-inertial reference systems, i.e., those that have a relative acceleration with respect to each other. (In the theory of special relativity, reference systems that have only a constant relative speed with respect to each other are taken into account.) In this way space-time becomes a physical entity, which is no longer indifferent to physical phenomena, but is modified by the presence of massive bodies or the presence of energy and acquires a curvature. So, gravity is no longer a force at a distance, but becomes a field theory (Weinberg 1972, 67–70). This means that, if I have a body of mass m_1 and move its position, another body of mass m_2 will feel the displacement (perturbation) of the position of m_1 not immediately, but after a time equal to the time it takes the light to travel the distance that separates the bodies m_1 and m_2. A consequence of all this is that if a ray of light that is emitted from a distant star to reach us passes close to the sun, it is deflected by the curvature generated by the mass of the sun, so that its apparent position with respect to an observer on earth does not coincide with its actual position.

Immediately after the publication of the theory of general relativity, many physicists and mathematicians tried to derive exact solutions from the equations to which it gave rise. Friedmann, Lemaître, Robertson and Walker (FLRW) (Weinberg 1972, 407–409), independently of each other, found that, if we assume that the distribution of matter in the universe is homogeneous and isotropic, on a large scale, the solutions of the equations of general relativity foresee a universe that is, in the spatial part, a three-dimensional surface of a four-dimensional sphere, whose radius represents time. This sphere representing the universe expands in time. To express this by analogy, three-dimensional space behaves as if it were a two-dimensional spherical surface on which all the galaxies and elements in general of the universe are located. Like a fairground balloon, this sphere expands, so the distance between galaxies increases over time.

3 The Birth of Big Bang Cosmology and the Concordism in Theology

Einstein did not like this solution, and he branded it as "mathematically" correct but "physically" wrong. For this reason, he modified the equations of general relativity, introducing a constant, called "the cosmological constant", which provided a solution of a static universe that did not expand (Weinberg 1972, 613–616). However, the measurement of the redshift of the spectral lines of the *nebulae* (galaxies) by Hubble proved the recession of all galaxies from an observer on Earth, demonstrating that the universe was actually expanding. When he realized this, Einstein declared that he had made the biggest mistake of his life.

But if the universe expands, then, going back in time, there must have been a primordial period when it was very small. This gave rise to the idea of the Belgian priest and cosmologist Georges Lemaître, who hypothesized that in the beginning the universe was the size of an atom (which he called "the original atom"), and that therefore the laws governing this original universe-atom were those of quantum mechanics.

Lemaître also had the distinction of having deduced the existence of the recession of the galaxies from the cosmological model FLRW, purely theoretically, before Hubble's measurement. However, as there were still no accurate measures of the velocities and photometric distances of the galaxies, he published his article in a little-known French scientific journal (International Astronomical Union

2018), so the credit for the famous law that bears his name went to Hubble (Lambert 2016, 121–145).

This view of the evolution of the universe aroused much suspicion among many scientists, who noticed a close proximity to the biblical episode of creation in the Book of Genesis. To make fun of Lemaître's theory, Fred Hoyle, an English astrophysicist, called this theory the "Big Bang". He developed his own theory called the Steady State Universe, in which the universe expanded while maintaining a constant density of energy-matter, so that it had no beginning and no end, but it was necessary to assume a continuous production of matter-energy (Weinberg 1972, 459–464).

These two models of the universe remained in competition with each other for several years. On 22 November 1951, Pius XII – who was certainly one of the pontiffs most attentive to scientific questions – delivered a speech at the Pontifical Academy of Sciences. It was entitled "Un'ora" (An hour) (Pope Pius XII 1951) and in it he hinted that the cosmological model of the Big Bang confirmed the story of the creation of the world in the Book of Genesis. Of course, Pius XII was a very intelligent person and highlighted that more philosophical as well as theological investigations were needed to arrive to the conclusion that the Big Bang theory and the *Fiat Lux* in *Genesis* were the same thing. But he was convinced, as he wrote it in his speech, that an enlightened mind could have made a jump (*a Pindaric flight*) and recognized traces of the God of creation and Love in the scientific description of the beginning and evolution of the universe.

In this speech, with its clearly Neo-Thomistic approach, the pope re-proposed the "ways" of establishing the existence of God of Saint Thomas Aquinas, especially the first and the fifth, based, respectively, on mutability and finality. In this Neo-Thomistic approach, he employed, to support the mutability in the world (St. Thomas used the terms "generation" and "corruption" to indicate this mutability in the world), that there are continuous changes in physical and chemical phenomena. On the other side, to support the finality in the world, he used the second principle of thermodynamics, according to which the entropy of a closed physical system always increases. Therefore, increase of entropy, in the physical world, was evidence of a teleology. This theological approach, in which scientific theories were used to confirm theological positions, was later renamed "concordism".

Lemaître felt called into question by this speech, because in the past he had already been suspected of concordism. Moreover, a problem then arose, because the following year the meeting of the International Astronomical Union (IAU) was to be held in Rome, and Pius XII had been invited to give the inaugural speech. Therefore, Lemaître, in his way to South Africa stopped in Rome, where, through the mediation of the Jesuit Fr. O'Connell, the then director of the Vatican Observatory, he met Pius XII. Obviously, we do not know the content of the conversation between the pope and Lemaître. The fact is that Pius XII gave his inaugural speech at the IAU on 7 September 1952, but made no mention of concordism (Turek 1989, 167–175).

For his part, Lemaître continued to always keep the theological and scientific planes distinct, as two parallel planes that do not intersect or, better still as two independent sources of knowledge. In 1965 two scientists from Bell Laboratories, Penzias, and Wilson, thanks to a large microwave antenna built for communication, detected a uniform radiation in all directions with a temperature of about 3 degrees Kelvin. This radiation, now known as the "Cosmic Microwave Background" (CMB), represents the first light emitted by the universe 400,000 years after the Big Bang, and can only be explained by the Big Bang theory, and not by that of the "Steady State Universe" (Weinberg 1972, 511).

Today the scientific community agrees that the universe in which we live was born 13.82 billion years ago, from a very hot phase, involving a cosmological event we call the "Big Bang". In the initial moment, called "the singularity", Einstein's equations are no longer valid. Immediately afterward, the universe underwent a great expansion, at a much greater rate than it is expanding now, an exponential expansion known as "inflation". About 400,000 years after the Big Bang, the universe emitted its first light, and then, little by little, all the structures were formed. In 1998, the study of the redshift of the light spectrum from Type IA supernovas showed that the universe was not only expanding, but also accelerating. Now, if the force responsible for this expansion is gravity alone, then the universe should expand by decelerating. If it accelerates, it means that a force opposite to gravity, a sort of anti-gravity, is operating. To explain this acceleration, the cosmological constant that Einstein had introduced in his equations was taken up again and the hypothesis of the existence of a non-visible energy called "dark energy" was formulated. In this way one gets a system that explains an accelerated expansion of the universe (Weinberg 2008, 1–100).

The nature of "dark energy" is not yet clear, and moreover it has not yet been directly observed. According to the latest measurements provided by the Planck satellite, "dark energy" should account for 68.3% of all energy-matter in the universe (Planck Collaboration 2021, 1–3). To this "exotic" element from the point of view of observations is also added "dark matter". In fact, the rotation curves of spiral galaxies have a radial velocity graph as a function of the distance from the center of the galaxy that does not coincide with the theoretical graph, which can be explained by the presence within the galaxy of non-conventional matter, which is called "dark matter". The latter is 26.8% of the total matter-energy of the universe, while the matter-energy observed in the universe is only 4.9%. It is therefore understood that this model of the universe, called ΛCDM (Λ is the cosmological constant and refers to dark energy; CDM stands for *Cold Dark Matter*, not high-energy dark matter), has many aspects that are still research topics and that do not allow us to say that a definitive model has been established (Planck Collaboration 2016, 1–31).

4 Quantum Gravity and Some Questions Involving Science and Faith

Now we need to delve into the "Quantum Gravity" phase of our universe; what Lemaître had called the "primitive atom", because it has given rise to many debates on questions of science and faith.

As it is usually classified, quantum gravity is a phase of our universe that goes from the initial instant to Planck time, which is about 10^{-43} s. It is a very small interval of time, in which Einstein's equations, which we mentioned above, lose their predictive meaning. Therefore, we need a new theory that combines two worlds of physics that seem irreconcilable: quantum mechanics, which provides the laws of physics for the behavior of particles at the atomic and subatomic levels, and Einstein's General Relativity, which describes the behavior of bodies on very large scales, beyond the galactic scales. This theory, which "should" – the conditional here is necessary, since we do not yet have a definitive theory – combine general relativity and quantum mechanics is called "quantum gravity".

One of the first approaches to this theory is the so-called "canonical approach", which basically consists in the attempt to write an equation for the wave function that should indicate the entire primordial universe. This is Wheeler-DeWitt's equation (Kolb and Turner 1994, 447–464) and is without the variable "time", so it is said that the wave function of the primordial universe is timeless. This has generated a lot of confusion; however, it should be noted that a parameter of evolution is still necessary to describe the evolution of the universe: for example, in some cases, as the universe expands, the volume of the universe is used as a parameter of evolution.

Hartle and Hawking have come up with a solution for the Wheeler-DeWitt equation that goes under the name "Hartle-Hawking proposal". It is a fairly complex solution from a mathematical point of view, which aims to eliminate the problem of the initial "singularity". The Hartle-Hawking model suggests a sort of phase transition, in Planck time, from the Lorentz regime to the Riemannian regime. Thus, under Planck time, there are compact surfaces that have no singularity, and therefore no privileged points. For this reason, as Hawking has repeated in public lectures and in several of his writings, there is no beginning and there is no need for a God who acts as a "first cause" to initiate the process through which the universe evolves. Hartle and Hawking claim that, under Planck time, time is imaginary and therefore behaves like other spatial coordinates. The transition from phase to Planck time makes the transition from imaginary time to the physical time of true evolution. This "Riemannian phase" of the universe, which goes from the initial instant of the universe to Planck time, is the "Vacuum state" of the Hartle-Hawking model.

Hawking maintains that the Vacuum state is the *nihil* of the doctrine of *creatio ex nihilo*, and that the imaginary time under Planck time would explain the absence of time "required" by *creatio ex nihilo*. As the Jesuit William Stoeger (Stoeger 2009, 174–181) has pointed out, excessive manipulation is being carried out here by Hawking. The *nihil* in the above-mentioned doctrine means really "nothing", the absence of everything even of energy. Nothing existed; even the physical laws were absent. While, in reality, in this quantum vacuum, there exists both energy and the physical laws that regulate phenomena. Moreover, to say that time does not exist in the sub-Planckian region, because time is imaginary, is also far-fetched.

The problem is that the Big Bang and "singularity" refer to an original event whose cause is unknown, and scientists fear that this cause must be a God who, like a demiurge, gives birth to the universe and then disappears, as suggested by some forms of deism. For this reason, Hawking felt the need to develop a model of quantum gravity that is completely autonomous and does not need to resort to an original cause: that is, according to him, one can do without God.

However, there are two points to be clarified here. The first is that the Hartle-Hawking model is not *the* fundamental solution to quantum gravity, but *a* possible solution, which we do not even know whether it has occurred in nature (Stoeger 2009, 174–181). The second is that to think that one has to resort to a God-demiurge to explain a cause one cannot clarify otherwise is to commit a philosophical error. Descartes made a similar error when he resorted to the existence of a good God to make sure that no one (the famous argument of Descarte's devil) had deceived him when he was building his philosophical system. This God who is used when something cannot be explained is called by some the "*God of the gaps*".

But this is not a proper way of thinking in theology. In fact, if one day it were discovered that there is a phase of the universe before the Big Bang – and pre-Big Bang theories already exist – then this God-demiurge would no longer be useful, because science would explain that there is something before the Big Bang, and therefore God would not "useful" anymore and then it does not exist.

However, the problem of the beginning of the universe continues to be present in the minds of many scientists, because it is considered as a "prime cause" that needs recourse to a God-demiurge, especially since this beginning is confused with the term "creation".

The Christian concept of creation is, instead, completely different from that of the God-demiurge of scientists. First, God creates from a state where before there was really nothing (*creatio ex nihilo*), i.e., neither initial energy nor physical laws. Indeed, he creates both energy and physical laws from nothing and keeps them in existence. God creates the world and all its creatures into *being*. Creating creatures into *being*, he reduces to zero the distance between Him and them. He supports his own creation sustaining the creatures continuously into being (*creatio continua*). Creation is then a "relationship", as Saint Thomas Aquinas said or *creatio est relatio* (Aquinas, *Summa Theologiae*, I, q. 45, a. 3), between the Father and the Son, who is the "Logos" through whom the Father creates the world and thanks to whom creation has a "logical" structure. This relationship between the Father and the Son is a relationship of Love, that is, the Holy Spirit, the third Person of the Trinity. In this way we have the *creatio ex amore* (Youngs 2014, 165–186), so in creation we find the traces of God's Love. Therefore, creation has a basically Trinitarian structure (Gamberini 2007, 148–163). There is a complementarity between Big Bang cosmology and *creatio ex nihilo.* Big Bang cosmology addresses, naturally, a question on creation that it cannot deal inside scientific categories. The doctrine of *creatio ex nihilo* answers, from a catholic theological perspective, the question on creation. Big Bang cosmology and *creatio ex nihilo* can live together without conflicts if we accept, following Lemaître intuition, they are two different and distinct *magisteria.*

5 Conclusions

In this essay we have summarized the temporal outline of our universe. First, we have described the evolution of its physical elements starting from the Big Bang up to the formation of its "cosmological" structures (galaxies and clusters of galaxies) in the universe. Also, and in a parallel way, we have mentioned the history of the fundamental interactions in our universe. We have introduced the Planck era, the Grand Unified Theory (GUT), the strong and electroweak era up to nowadays four fundamental interactions. We have explained the basic elements of Einstein's General Relativity and modern relativistic cosmology. Then we have briefly mentioned that the FLRW homogeneous and isotropic cosmological model plus dark energy and dark matter constitute the ΛCDM standard model in cosmology. The history of the Big Bang theory has been exposed with particular emphasis on Mons George Lemaître and the problem of "concordism" in theology. We have stressed that, after a "concordist" phase in his personal life, Lemaître always kept science and faith very distinct. Contemporary quantum cosmology model, as the Hartle-Hawking no-boundary boundary proposal, has been discussed. We have highlighted that any theological claim on the *creatio ex nihilo* based on theoretical physics considerations could be only completely misleading. We have briefly summarized what is *creatio ex nihilo* from pure theological considerations.

Today we see that the ΛCDM standard cosmological model works quite well with observational data; however, as we have explained, it is necessary to use ad hoc "elements", such as dark matter and dark energy, to explain some otherwise unexplained phenomena. In this sense one could think, with all the reservations and cautions of the case, that there could be an analogy between the theory of the epicycles of Ptolemy's geocentric system, invented to explain the motion of the planets, and the hypotheses of dark matter and dark energy, introduced to adapt the cosmological model to otherwise unexplained phenomena. In other words, it must be said that, despite all the progress that has been made in science, and in particular in current cosmology, the myth of a "very precise" science, without any shadow, must certainly be debunked. The truth, however, is that even the scientific models that we possess today and which we use to describe nature have limitations, and therefore do not possess to any degree the character of infallibility that a new dogmatic "scientism" would like to attribute to them.

Since ancient times there has always existed a tight connection between cosmology and religion. In ancient cultures, starting from the harmony and order existing in the visible universe – which at that time was simply the starry sky – people have always tried to hypothesize the existence of an "architect" God which is the cause of this harmony. Let us remember, as one out of many, the so-called "cosmological proofs of the existence of God", where from the "contingency" of the world philosophical arguments deduced the necessity of the existence of a first cause, God, Who is also the guarantor of universe harmony. However, the modern conflicts – for example, the "Galileo case" and the subsequent fracture between science and theology – lead us to think that, following Lemaître, the right approach, in the science-theology debate, is the separation between the theological and scientific planes or *magisteria*. But this does not prevent a mind, enlightened by the Grace of God as Pius XII was mentioning in his 1951 speech, from seeing in the harmony and order of the universe a beauty that reflects the imprint of the Creator and the Love with which God created and wove the universe. However, this is not proof of the existence of God, but rather an *a posteriori* observation, valid only for those who are either already believers or accepting God's Grace to believe.

References

Aquinas, Thomas. *Summa Theologiae*. South Bend, IN. Christian Classics, Notre Dome University.

Gamberini, Paolo. 2007. *Un Dio relazione* [*A Relational God*]. Rome: Città Nuova.

Gibney, Elisabeth. 2018. "Belgian Priest Recognized in Hubble-Law Name Change". *Nature*. http://dx.doi.org/10.1038/d41586-018-07234-y.

Kolb, Edward W., and Michael S. Turner. 1994. *The Early Universe*. New York: Addison Wesley.

International Astronomical Union. 2018. "Thirtieth General Assembly: Resolution Presented to the XXXth General Assembly". *Resolution B4* (October).

Lambert, Dominique. 2016. *The Atom of the Universe*. Krakow: Copernicus Center Press.

Pope Pius XII. 1951. "Discorso di Sua santità Pio XII ai cardinali, ai legati delle nazioni estere e ai soci della pontificia accademia delle scienze". https://www.vatican.va/content/pius-xii/it/speeches/1951/documents/hf_p-xii_spe_19511122_di-serena.html, last accessed 19 September 2023.

Planck Collaboration. 2016. "Planck 2015 Results XIV. Dark Energy and Modified Gravity". *Astronomy and Astrophysic* 594 A14: 1–31. DOI: 10.1051/0004-6361/201525814.

Planck Collaboration. 2021. "Planck 2018 Cosmological Parameters (Corrigendum)". *Astronomy & Astrophysics* 652, C14: 1–3. DOI: 10.1051/0004-6361/201833910e.

Stoeger, William R. 2009. "Is Big-Bang Cosmology in Conflict with Creation?" In *The Heavens Proclaim*, edited by Guy Consolmagno, 174–181. Vatican City: Vatican Observatory Publication.

Stoeger, William R. 2010. "The Big-Bang, Quantum Cosmology and *creatio ex nihilo*". In *Creation and the God of Abraham*, edited by David B. Burrell, Carlo Cogliati, Janet M. Soskice, and William R. Stoeger, 152–175. Cambridge: Cambridge University Press.

Turek, Jozéf. 1989. *Georges Lemaître and the Pontifical Academy of Sciences*. Vatican City: Specola Vaticana.

Weinberg, Steven. 1972. *Gravitation and Cosmology*. New York: John Wiley and Son.

Weinberg, Steven. 2008. *Cosmology*. Oxford: Oxford University Press.

Youngs, Samuel. 2014. "Creatio Ex Amore Dei: Creation out of Nothing and God's Relational Nature". *The Asbury Journal* 69: 165–186. https://place.asburyseminary.edu/asburyjournal/vol69/iss2/12, last accessed 19 September 2023.

Alberto Peratoner

“Qu’est-ce qu’un homme, dans l’infini?”. Eternity and Infinity in Blaise Pascal and in the 17th-Century Geometrizing Ontologies

Abstract: At the rise of Modernity, in the wake of Renaissance Platonism, with Cusanus and, later, Giordano Bruno, the notion of infinity assumes a central role in understanding the structure of the Universe, in defining the relationship between God and the World, to assume an even greater importance in the geometricizing ontologies of the 17th century. Thus, if in Descartes and, subsequently, in authors such as Spinoza, Malebranche and Leibniz, the finite – infinite relationship outlines the field of man’s condition in relation to nature and God, in Pascal it defines the space of an onto-anthropological understanding of existential calibre, where man finds himself suspended in a state of finitude between the infinitely large and the infinitely small, in which he experiences a fluctuating instability in the tensions of the *contrariétés* of *misère* and *grandeur*, and opens up the consideration of the finiteness of the existence temporality facing the perspective of eternity, as conceived by the Christian faith, in God.

At the dawn of Modernity, in the wake of Renaissance Platonism, with Cusanus and, later, Giordano Bruno, the notion of infinity took on a central role in the comprehension of the Universe’s structure, in the definition of the relationship between God and the World. It also confirms itself by this path as a notion analogically related to that of eternity, which, in the development of the mathematical interpretation of cosmic reality, tends somehow to reflect its geometrization. Thus, the two fundamental empirical coordinates, space and time, transposed to transcendence, still come to subsist in their infinite and eternal dilatation, of an infinitized eternity, and an eternalized infinity.

In Cusanus, every entity of created nature, as a particular determination of being, is conceived as a finitization of the exemplar present in divine thought, and that in God is merged into the infinite that he is in his perfection, whereby the work of creation responds to a logic of de-limitation that deploys in the constitutive *explicatio* of the finite multiple of natural reality, and hence in the *contractio* that every entity of nature congeals within itself.

https://doi.org/10.1515/9783111313610-018

With Giordano Bruno, the infinity of God, echoed by Cusanus even in the conception of the *coincidentia oppositorum*, is extended to the very reality of the world, which of God is considered the expression, as much as, reciprocally, God himself is nature, in its productive force – *natura naturans* –, so that the infinity that unites them becomes the reason for the eternity of both.

1 Eternity as a Temporal Infinity. Some Theoretical Perspectives of 17th-Century Geometrizing Ontologies

The reflection about infinity later comes to assume even greater relevance when new developments in mathematics concur to form what we may define as the *geometrizing ontologies* of the 17th century. By this we mean those ontological-metaphysical conceptions, mainly characterized by a systematic and holistic theoretical framework, in which the mathematical sciences, especially in their "geometric" expression, assumes the paradigmatic form of a rationality of the highest efficacy on the methodological level as well as on that of the effective representation or rendering of the Whole of Being sense. The highest expression of this epochal tendency can be considered the Leibnizian project to establish a *characteristica universalis* as the cipher of a mathematical language capable of describing every feature of reality.

The increased interest for the logical-mathematical, and at the same time ontological, statute of infinity is thus inscribed within this framework of a tendential geometrization of the philosophical comprehension of being, where the reflection on mathematical infinity developed by Galileo converges. The latter's contribution in effect consists in the demonstration of the incommensurability of the notion of infinity in relation to finite quantities, gained through the bi-univocal correspondence between the series of natural numbers and the set of their squares, which leads to the paradox – known as that of Galileo, and described in the *Discorsi e dimostrazioni matematiche intorno a due nuove scienze* (1638) – whereby it is possible to find as many elements for infinite sets as for a subset of its own, which is impossible for finite sets, where stands the principle that "the whole is always greater than the part".

Descartes takes up the identification of the infinite with God only, and detaches from it the notion of indefinite, to express the quantitative magnitudes and "of possible things" (Descartes 1644, I). He also makes the relationship between the finite and the infinite an expression of the ontological difference between man and God, echoing an ancient Augustinian theme that leads him to reformulate, in the third of the *Meditationes de prima philosophia*, the proof *a contingentia hominis* of God's existence. This, in fact, relies on the presence in human conscience of the idea of the infinite as not derivable by a mere increment of the finite.

> In fact, even if the idea of substance were in me by the very fact that I am substance, not for this, however, would there be the idea of the infinite substance, since I am finite, unless it came from a substance which was infinite in reality. And I must not consider that I do not perceive the infinite by a true idea, but only through the negation of the finite, in the same way as I perceive stillness and darkness by the negation of motion and light; for, on the contrary, I manifestly understand that there is more reality in infinite substance than in finite one, and that, therefore, in me, the perception of the infinite comes before that of the finite, that is, that of God before of myself. In which way, indeed, would I intend to doubt, to desire, that is, that I lack something and am not entirely perfect, if there were in me no idea of a more perfect entity, comparing myself to which I could recognise my defects? (Descartes 1644, III)[1]

On its side, the gnoseological level of a potential indefinite increase in knowledge emphasizes and confirms the substantial ontological difference and the unfillability of the infinite by an increase of the finite.

> Moreover, even if my knowledge should always increase, I nevertheless intend that it will never become infinite in act, because it will never reach the point not to be susceptible to even greater increase; instead, I judge that God is so infinite in act that nothing can be added to his perfection (Descartes 1644, III).

The infinity of Spinoza's God deploys itself in *natura naturata* as the expression of his unlimited generative power of *Natura naturans*, as "an absolutely infinite entity, that is, the substance consisting of infinite attributes, each of which expresses eternal and infinite essence" (Spinoza 1677, I). Proceeding from the necessity of God's nature, of His infinite attributes and of all their modes, given the unity of the Substance that is God Himself and God alone, *natura naturata* retains the same infinity, since attributes and modes "are considered as things that are in God, and which without God can neither be nor be conceived" (Spinoza 1677, I).

In Malebranche, on the contrary, the onto-axiological difference between infinite and finite is pushed as far as the reduction to nothingness of the second compared to the first and to the total extraneity and impossibility of a relationship

1 On the question of infinity in the 17th century, from different perspectives that also outline the common relationship of different authors, in this regard, with Descartes and Cartesianism as their focal reference point, see Lardic 1999.

between the two dimensions, which motivates the retranslation of their correspondence in the terms of occasionalism: “The finite, however great it may be, cannot establish in itself any relationship with the infinite. Ten thousand centuries in relation to eternity is nothing. The relation of the extension of the entire universe to spaces having no limits cannot be expressed other than through zero” (Malebranche 1684, II, VII).[2]

The same act of creation itself, in the ontological dependence of the finite which comes to realize in space-time determinacy, expresses its nothingness, while, within the circumscribed sphere of finiteness, greater and lesser degrees of perfection of created beings are recognizable, and among these the excellence of man emerges, insofar as he is the keeper of an idea of the infinitely endless infinite which enables him to some relationship with it.

> It is manifest, indeed, that the spirit perceives the infinite while not comprehending it, and that it has a very distinct idea of God which can come to it only from the union it has with him; for it cannot be conceived that the idea of an infinitely perfect being, such as we have of God, should be something created. But not only does the spirit have the idea of the infinite; it has it even before that of the finite. For we conceive infinite being only insofar as we conceive being, without thinking whether it is finite or infinite. [. . .] Thus the spirit perceives nothing except the idea it has of the infinite (Malebranche 1674, III, II, 6).

Leibniz, in order to save the reality of the finite as such and precisely distinct from the infinity of God, as well as the fundamental unity of being, affirms the reality of the actual infinity by extending it to nature, but he distinguishes in it the maximum infinity, proper to the absolute, and the infinity relative to the maximality proper to the singular genera, thus space in its totality as the maximum of every extension and eternity with respect to temporal succession. The monadic structure of reality, then, presents in the lowest degree of the “metaphysical points” a contraction and mirroring of the infinity of the Universe in the infinity of its own relations and perceptions (Leibniz 1720, 36–37). In its turn, the created Universe presents itself as an infinity in a finite mode, since it is individually determined, elected by God to real existence as chosen from an infinity of possible worlds (Leibniz 1710, I, 7–9).

Thus, in Descartes and, later, in authors such as Spinoza, Malebranche and Leibniz, the finite-infinite relationship variously delineates the field of man’s condition in relation to nature and God, where the analogical correlation between infinity and eternity as expressions of the Absolute is clear, in relation to the two forms of the world’s ontic finitude, that is of spatial finiteness and temporality.

2 About infinity in Malebranche, see Carbone 2004.

Now, with respect to these conceptions, in Pascal the ontological sense of the infinite in relation to the finite, derived from the mathematical sciences, defines the space of an onto-anthropological comprehension in an existential perspective, where man finds himself suspended in an intermediate finitude state between the infinitely large and the infinitely small, in which he experiences the fluctuating instability within the tensions of the *contrariétés* of *misère* and *grandeur*, and opens to the consideration of the temporal finitude of existence in face of the prospect of eternity in God as conceived by Christian faith.

2 Pascal's Mathematical-Geometric Double Infinity and His Three-Order Ontology

When Victor Cousin stumbled on the Pascal's *pensée* "The eternal silence of these infinite spaces frightens me" (Pascal 2000),[3] he wondered if "this sinister phrase", isolated from the rest, was not "like a mournful cry suddenly emerging from the abysses of the soul, in the desert of a godless world" (Cousin 1842, 546). Cousin was at the time among the most convinced promoters of Pascal's *légende sceptique* (an interpretation that conveyed the romantic image of a tormented and *pyrrhonien* Pascal, unable to find a solution in his own search for the reasons for the credibility of Christian faith), and neither the state of research on the *Pensées* allowed at the time to have an idea of the structure of Pascal's unfinished Apology and consequently a precise contextual framework to which refer the meaning of the single fragments (Scherer 1885, 181; Peratoner 2002; Peratoner 2011).[4]

The brief thought that shocked the Cartesian philosopher Cousin plays indeed a strategic role in the design of Pascal's Apology, and at the same time expresses

3 "Le silence éternel de ces espaces infinis m'effraie". Pascal 2000, 15/187 (L 201; B 206). Quotations from the *Pensées* bear the numbering of the Le Guern edition (Paris 2000), with the number of the section (*liasse*) in which the fragment is located in the *Première Copie* (in Arabic numerals for the first 27 so-called *liasses titrées*, in Roman numerals for the subsequent units, without titles) separated by a slash from the following fragment number, followed by Lafuma (Paris 1951) and Brunschvicg (Paris 1897) numbering, in round brackets and preceded by the respective initials L and B. All quotations are provided in our own translation.

4 Still in 1885, in front of the same fragment, Edmond Scherer will declare, echoing the same interpretative line, that "the terror that is betrayed in these words is not simulated; Pascal has attempted to probe the infinite; he asked the immensity and the evolving worlds why the universe exists, and nothing answered him" (Scherer 1885, 181). About the philological question of the recoverability of the unfinished Apology plan and its consequences for the correct interpretation of Pascal's thought, see Peratoner 2002, I, 277–752; 791–808; Peratoner 2011, 167–220.

and urges the voice of conscience in the face of the immensity of space and time deployed before the eyes of mind. For our purposes, it is particularly interesting in that it associates the eternity of time with the infinity of the contemplated sidereal space: the infinity of space manifests itself in the terms of a silent eternity: eternity and infinity intertwine and refer back to each other, urging the conscience to the ultimate and deepest existential questions, even if lying under the mantle of the rationalist optimism of an emerging culture in which Pascal already perceives the premonitory signs of the scientistic and reductivist Enlightenment of the following century.

About Pascal's determination of the *méthode géométrique* as an epistemological paradigm, the most significant work is undoubtedly the large double fragment *De l'esprit géométrique*, written in 1655, in which we find important statements regarding the notion of infinity and its role on the mathematical-geometric as well as the metaphysical level.

In establishing the axiomatic method, Pascal reaches the existence of "primitive terms that cannot be further defined", and hence to theorize three orders of magnitudes, heterogeneous but sharing certain properties, indefinable because of their extreme evidence that also constitutes their perfection: motion, number, space (Pascal 1962–, III, 401).[5]

Now, the main common property of these orders of magnitude consists in the "two infinities occurring in all things, one of greatness, the other of smallness" (Pascal 1962–, III, [1,22], 402). Pascal dwells on exemplifying this double infinity by setting the concept of time beside the three orders:

> The same applies to time. [. . .] whatever the movement, the number, the space, the time, there is always one greater and one smaller, so that they all sustain themselves between nothingness and infinity, being always infinitely distant from these extremes (Pascal 1962–, III, [1, 23], 402–403).

Space and time are thus both governed by the tensional structure of a double infinity, whose maximal expression represents, in parallel, spatial infinity and eternity in the sense, here, of infinite duration. Nearing the conclusion of the first fragment section of the *Esprit géométrique*, Pascal comes back to the *deux infinis*, which, "although infinitely different, are nevertheless related to each other, in

---

5 See Pascal 1728, [1, 21], 401; "These three things, which comprise the entire universe, according to the words *Deus fecit omnia in pondere, in numero, et mensura* [see *Book of Wisdom*, 11:21], have a reciprocal and necessary connection. For one cannot imagine movement without something in motion, and since this thing is one, this unity is the origin of all numbers; finally, since movement cannot be without space, one sees these three things included in the first" (Pascal 1728, [1, 22], 401).

such a way that knowledge of the one necessarily leads to knowledge of the other" (Pascal 1962–, III, [1, 36] 410), and anticipates an interesting development that will concern the anthropological conception elaborated within the Apology argumentative path:

> Those who see such truths clearly will be able to admire the greatness of nature's power in this double infinity surrounding us on all sides and learn from this marvellous consideration to know themselves, seeing themselves placed between an infinity and a nothingness of extension, between an infinity and a nothingness of number, between an infinity and a nothingness of movement, between an infinity and a nothingness of time. On the basis of this we can learn to estimate ourselves at the right value, and form reflections more valuable than everything else in geometry (Pascal 1962–, III, [1, 39], 411).

The anthropological-existential development that will take on the reflection on the infinite in the *Pensées* is thus announced: in line with the general continuity of Pascal's thought from science to philosophy and theological reflection, it is already in the scientific setup of the treated questions that he recognizes, and anticipates, what he evidently considers a natural extension of the reflections developed in his scientific research. Here he suggests the transcending of the strictly geometrical horizon, a crossing made possible by the analogy expressed by this same perspective, which takes on a practical-existential significance in the anthropology of an "average" positioning of the personal human subject between the two extremities of the infinitely large and the infinitely small, primarily conceived on a mathematical basis.

The fragment *De l'esprit géométrique* concerns the notion of infinity also in relation to the contemporary provocations of Antoine Gombaud, Chevalier de Méré, about which we have an interesting testimony in a letter by him addressed to Pascal.[6] In this one, Méré, esteemed by Pascal for his *esprit de finesse* – meaning an intuitive spirit of rare fineness in the things regulating good living and human relationships –, reproaches the geometric method for conducting obscure and therefore suspect, even groundless, demonstrations, and lead to conclusions

---

6 Actually, the letter is a literary *fictio*, published by Méré himself along with other his letters in 1682, but faithful about a real controversy dating from 1654–1655. There is, moreover, a letter from Pascal to Fermat dated 29 July 1654, in which he mentions "a difficulty that surprised M . . . a lot, because he has a very fine spirit, but he is not a geometer at all. This is, as you know, a great defect. And even if he does not understand that a mathematical line is divisible to infinity and believes to understand that it is composed of joined points in a finite number, I have never been able to convince him" (Pascal 1962–, II, 1142). In his commentary on Pascal's letter, J. Mesnard conjectured that, having published Pascal's text in the collection *Varia opera mathematica D. Petri de Fermat* in 1679, Méré would have been moved to reply in a fictitious letter (Pascal 1962–, III, 349).

that are inconceivable to him, such as the infinite divisibility of space and the existence in nature of the infinitely small, which he finds "far away from common sense" (Pascal 1962–, III, 353, 354, 356).[7]

The question has resonance in *De l'esprit géométrique*, where Pascal dwells on the primality of first principles as not further subjectable to demonstration:

> Geometry cannot define objects nor prove principles; but for this one and advantageous reason, that the one and the other are in an extreme natural clarity that convinces reason more powerfully than speech. [. . .] There is no natural knowledge in man that precedes these and surpasses them in clarity. Nevertheless, in order that we may have an example of everything, we find spirits, excellent in all other things, who are stunned by these infinities, and who cannot in any way allow them (Pascal 1962–, III, [1, 24, 26], 403).

As in other cases, as a great treasurer of his own experience, Pascal takes advantage of even a matter of simple conversation like this, and makes it a motive to reflect on the vertigo the intellect itself feels in engaging with the thought of infinity, which, although it responds to logical incontestable principles, provokes in some a stunning effect, even to the repulsion of conceiving some of its essential properties. What follows clarifies the sense of the applicability of these principles to the design of the unfinished Apology:

> There is no geometer who does not believe space to be infinitely divisible. [. . .] And yet there is none who comprehends an infinite division; and we assure ourselves of this inconceivable truth by this reason alone, but which is certainly sufficient, that it is perfectly understood that it is false that by dividing a space one can arrive at an indivisible part, or that it has no extension at all (Pascal 1962–, III, [1, 27], 404).

The absurdity of this last hypothesis highlights the coexistence, in the conceiving of the infinite by the *géomètre*, of two contrasting dispositions in the conscience: the motivated and conscious acceptance of the infinite divisibility of space and at the same time its incomprehensibility. And indeed Pascal will take up, in the *Pensées*, with regard to the incomprehensibility of the main mysteries of the Christian faith, the sense of this precious acquisition, which dissociates for apologetic purposes the concept of incomprehensible from that of unexisting, with a clear reference to what is discussed in *De l'esprit géométrique*: "incomprehensible. Everything that is incomprehensible does not cease to exist. The infinite number, an infinite space equal to the finite" (Pascal 2000, 11/139 (L 149; B 430)).

---

7 Méré, in Pascal 1962–, III, 354. To the "long reasoning drawn out point by point" that "prevents one from first entering into the higher knowledges that never deceive", Méré opposes "the lively spirit and fine eyes" and the aptitude to penetrate reality through "common sense" (Méré, in Pascal 1962–, III, 353, 356, 354).

The theory of the heterogeneity of magnitude orders, enunciated in *De l'esprit géométrique*, is formulated more precisely in the short treatise *Potestatum numericarum summa*, belonging to the group of annexes to the *Traité du Triangle Arithmétique*, dating to summer 1654. The theoretical configuration of the question induced Strowski and Humbert to perceive a connection with the distinction of the *trois ordres* (*corps, esprits* et *charité* – i.e., respectively, the orders of *physical, intellectual* and *supernatural* realities) set forth in the *Pensées* and to conjecture a transposition from the mathematical domain to the philosophical-theological one, where the incommensurability of the *trois ordres,* crossing and informing the anthropological constitution itself and, hence, human experience, an incommensurability described in terms of an "infinite distance", explains the lack of increase of an order by the addition of elements from another (Humbert 1947; Strowski 1908).[8]

In the *Pensées* Pascal first establishes an analogy between two relations, in which it is the notion of infinity to give the sense of the incommensurability of the different reality orders: "The infinite distance from bodies to spirits figures the infinitely more infinite distance from spirits to charity, as it is supernatural" (Pascal 2000, 23/290 (L 308; B 793)).

There then follows an exposition, in orderly succession, of the three levels of reality – *corps, esprits, charité* – which are defined through the reason of their heterogeneity, to the last of which is hooked the apologetic application of the discourse: the order of magnitude of Christ's manifestation on earth is incommensurable and, therefore, unrecognizable and invisible if considered through the evaluation criteria of the lower orders. Pascal concludes by observing the impossibility to derive any act of charity from the lower orders, and thus to explain their existence, and crowns the sequence with the affirmation of the supernatural as an order infinitely superior to the others: "All bodies together and all spirits together and all their productions are not worth the slightest motion of charity. This is of an infinitely higher order" (Pascal 2000, 23/290 (L 308; B 793)).

8 "Let us apply to the science of man the principle according to which a magnitude does not increase if lower order quantities are added to it in the number we want, and we will arrive at one of Pascal's most important and profound ideas" (Strowski 1907–1908, II, 290). See also Humbert 1947, 178.

3 An Anthropology of Finitude as a Suspension Between Two Infinities, Facing Eternity

To awaken in man the consciousness of his own condition and position in the universe, Pascal derives from the *géométrie* the concept of infinity as a representation of reality that shows his condition as suspended between infinity and nothingness, and which comes to reinforce the anthropology of the *contrariétés* of *misère* and *grandeur* that would have occupied the first part of the unfinished Apology. In the fragment *Disproportion de l'homme*, which extends mathematics to the comprehension of physical reality, man's position in the universe is described in great oscillatory movements, increasing and decreasing, between the opposing abysses of the infinitely large and infinitely small:

> May man therefore contemplate the whole nature in its high and full majesty, may he turn his gaze away from the low objects surrounding him. [. . .] May man, returning to himself, consider what he is in relation to what exists, may he observe himself as misplaced, and from this little prison in which he finds himself lodged, I mean the universe, may he learn to estimate the earth, kingdoms, cities, things and himself at their proper value. What is a man, in infinity? But to present him with another equally astonishing prodigy, seek in what he knows the most delicate things, [. . .] He will perhaps think that this is the extreme smallness of nature. I want to show him a new abyss in there. I want to paint him not only the visible universe, but the immensity of nature that can be conceived within the limits of this compendium of an atom, that he may see there an infinity of universes [. . .] (Pascal 2000, 15/185 (L 199; B 72)).[9]

Hence a substantial reconsideration of the human condition:

> For, finally, what is man in nature? A nothing in relation to infinity, an all in relation to nothing, a medium between nothing and all, infinitely far from understanding the extremes; the end of things and their beginning are for him invincibly concealed in an impenetrable secret. Equally unable to see the nothingness from which he is drawn and the infinity in which he is engulfed (Pascal 2000, 15/185 (L 199; B 72)).[10]

To express its ungraspability, the infinitely large is formulated as "an infinite sphere whose centre is everywhere, its circumference nowhere", in which imagination loses itself. Nor is this state of suspension between the two infinities limited only to the cosmological position in a spatial infinite, since Pascal specifies in the same text that "our intelligence occupies the same place in the order of intelligible things as our body does in the extension of nature", and in this way it thus

---

9 Emphasis ours.

10 About this oscillation between the two infinities, see Davidson 1988, 80.

comes to weld with the mentioned dialectic of *contrariétés* and, thus, to assume the existential dramaticity that demands a response.

A further consideration seems to relate to the *pari* argument, the well-known Pascal's wager: "In the presence of these infinities, all finites are equals and I do not see why to fix our imagination on one rather than another" (Pascal 2000, 15/185 (L 199; B 72)).

In effect, the *pari* fragment bears the heading *Infini – rien*, and the argument is precisely based on the comparison between finitude and infinity in which the estimation of gains and losses shows the first as ephemeral and unsatisfying for man's aspirations. *Infinity*, here, is referred to the perspective of *eternity* offered by Christian faith in relation to the limited duration and transitoriness of earthly existence. The tension between the *deux infinis* thus sets man before *infinity as eternity*. Of God and of himself as in God saved and eternally received.

This "third infinity", so to say, goes beyond the merely quantitative notion of eternity as an infinite duration or temporality, since what is at stake in Pascal's wager, in full consonance with the perspective of Christian Revelation, is that of an eternity as an intensive experience of qualitative infinity submerged in God's life. So, he envisages an infinity in which the quantitative is transfigured into a qualitative dimension and thus is the wager's object, since it is a matter of gaining "an infinity of infinitely happy life" (*une infinité de vie infiniment heureuse*) (Pascal 2000, 15/185 (L 199; B 72)).

Facing the eternity (the eternal infinity) of God, in his finite condition suspended between the double cosmic infinity, man realizes, rethinking the classical (and Cartesian) argument *a contingentia hominis*: "I feel that I might not have existed, since the self consists in my thought [. . .] therefore I am not a necessary being. I am not eternal nor infinite, but I clearly see that there is in nature a necessary, eternal and infinite being" (Pascal 2000, 8/125 (L 135; B 469)).

References

Carbone, Raffaele. 2004. "Gli infiniti e il finito in Malebranche: dallo smarrimento alla funzionalità della mente". *Annali della Facoltà di Lettere e Filosofia dell'Università degli Studi di Milano*, 57/1: 71–110.

Cousin, Victor. 1842. "Des Pensées de Pascal. Rapport à l'Académie française sur la nécessité d'une nouvelle édition des Pensées de Pascal". *Journal des Savants* [178]: 243–252; 333–358; 406–426; 490–505; 532–553; 608–625; 678–691.

Davidson, Hugh M. 1988. *Le concept de l'infini dans les* Pensées *de Pascal.* In *Pascal. Thématique des "Pensées"*. Edited by L.M. Heller and J.M. Richmond: Paris, J. Vrin: 79–91.

Descartes, René. 1644. *Principia philosophiae*. Amsterdam: L. Elzevirium.

Humbert, Pierre. 1947. *Cet effrayant génie . . . L'Oeuvre scientifique de Blaise Pascal*. Paris: Albin Michel.
Lardic, Jean-Marie. 1999. *L'infini entre science et religion au XVIIe siècle*. Paris: Vrin.
Leibniz, Gottfried Wilhelm. 1710. *Essais de théodicée sur la bonté de Dieu, la liberté de l'homme et l'origine du mal*. Amsterdam: I. Troyel.
Leibniz, Gottfried Wilhelm. 1720. *Lehrsätze über die Monadologie*. Frankfurt/Leipzig: Johann Meyer.
Malebranche, Nicolas. 1674. *De la recherche de la verité. Où l'on traite de la nature de l'esprit de l'homme, et de l'usage qu'il en doit faire pour éviter l'erreur dans les sciences*. Paris: A. Pralard.
Malebranche, Nicolas. 1684. *Traité de morale*. Rotterdam: R. Leers.
Pascal, Blaise. 2000. *Pensées*, in *Oeuvres complètes*. Edited by M. Le Guern. Paris: Gallimard, II, 541–1051.
Pascal, Blaise. 1962–. *Oeuvres complètes*. Edited by J. Mesnard. 4 Vols. Paris: Desclée.
Peratoner, Alberto. 2002. *Blaise Pascal. Ragione, Rivelazione e fondazione dell'etica. Il percorso dell'Apologie*. 2 Vols. Venice: Cafoscarina.
Peratoner, Alberto. 2011. *Pascal*. Rome: Carocci.
Scherer, Edmond. 1885. "La religion de Pascal". *Études Littéraires* 9: 175–194.
Strowski, Fortunat. 1908. *Pascal et son temps*. 3 Vols. Paris: Plon.
Spinoza, Baruch. 1677. *Ethica*. In *Opera Posthuma*. Amsterdam: J. Rieuwertsz, 1–264.

Leopoldo Sandonà

Eternity and Otherness from the Perspective of Dialogic Thinking. Inspirations and Contaminations in and from Romano Guardini, Franz Rosenzweig, and Nishida Kitarō

Abstract: This intervention attempts to set up a critical comparison with respect to the proposed theme – the relation between time and eternity – within the context of dialogical thought, meaning by this expression a varied constellation of authors in contemporary philosophy, both Jewish and Christian, philosophers and theologians. In particular, we will consider the Christian philosopher and theologian Romano Guardini (1885–1968) and the German-Jewish thinker Franz Rosenzweig (1886–1929). In both cases, although with different nuances, eternity is seen as openness in relationships (divine and human), an open semantic space neither not reducible to time, nor indifferent or alternative to it. In the final section, these two positions are compared with those of the Japanese thinkers Nishida Kitarō (1870–1945) and Nishitani Keiji (1900–1990).

This essay approaches the topic of the relation between time and eternity from the innovative perspective of dialogic thought, a varied and unsystematic constellation of authors crossing contemporary philosophy like a hidden river that borders on Jewish and Christian thinking, as well as philosophy and theology, and flourished in the 20th century (Zucal 2004; Casper 2022).

After a short digression about the etymological sense of verb the "to consider" (from the Latin *cum-siderare*) in Section 1, in the next two sections (Sections 2 and 3) I will expose and compare the two synergic philosophical positions of the Christian thinker Romano Guardini (1885–1968) and the Jewish thinker Franz Rosenzweig (1886–1929). In both cases, although with different accents, eternity is seen as openness to, and within, relationship (divine and human), an open space that is not reducible to time, yet is neither indifferent nor alternative to it. In Section 4, I will show the analogies between European dialogic thinking and the works of the Japanese philosophers Nishida Kitarō (1870–1945) and Nishitani Keiji (1900–1990). This comparison provides an excellent example of contamination between contemporary philosophy and theological reflection, though I will leave aside eschatological considerations, despite their closeness with the main topic.

https://doi.org/10.1515/9783111313610-019

1 Introduction: *Cum-siderare*

At a first glance, the reflections found in this introduction – inspired by the ancient etymology of the verb "to consider", from the Latin *cum-siderare* – may seem a detour from the main topic of this essay, namely the relation between eternity and time from the perspective of dialogic thinking. However, as "to consider" means "scrutinising with attention the stars [*sidera*]", the ideas rolled down here can be small trail markers steering us along a journey that will allow us to look at reality, Otherness, and God with focused and caring attention. Along this journey there is no shortage of biblical images that may help us. From its mention of the stars in its etymology, "to consider" is to be crossed by the surprise of an announcement, as it occurs to the shepherds, but also to set out to confirm omens and intuitions, like the Three Kings, knowing how to scan, observe, discern reality. In all these cases, existence looks beyond its immediateness, is not bent on itself, analytically lost in the labyrinth of data, but looks up, prospectively saved in the dimension of lived experience.

Perhaps our increasingly accurate technological tools have made us forget how to gaze in attention. I am not suggesting we should renounce to use modern telescopes – even less in Padua, celebrating the 800 years of the University that counts the work of Galileo Galilei among its most glorious pages. Yet we should not resort to technology to retrieve our lost attentive gaze on reality, in a sort of epistemological Luddism, either: we should rather refine and raise our gaze to the height of the tools we possess, while not renouncing to them. Today, even the quality level of medical care is measured by its ability to be in line with technology. We need not to decrease our technology to be more capable of care: we need accuracy – and appropriateness – in the alliance of the anthropological and the technological, the ethical and the technical. Neglecting care, even where it is not *per se* the outcome of a negative intention, leads to indifference and, in the long run, marginality.

After this detour (yet sometimes taking the wrong paths allows us to enjoy the most beautiful views), I come to question our main topic, the relation between eternity and time, two authors who spent a great deal on reflecting about dialogic thought: Romano Guardini and Franz Rosenzweig. A Catholic and a Jew, they both provided open and original readings of the venerable notion of eternity by their common resorting to philosophy and theology alike, where the latter provided a continuous re-translation of the given account of tradition.

Finally, this Jewish-Christian, philosophical-theological comparison, will also allow us to dare a wider West-East one.

2 "We are Time". *The Last Things* (1940) by Romano Guardini

The first dialogic author questioned here is Romano Guardini. In his short text *The Last Things*, dating back to Advent 1940, he interestingly turns to the last things in the midst of the catastrophe of the war, as if to affirm the impossibility for history to save itself, and the necessity to transcend forces that are merely human and worldly. The context is also telling, recalling not only the wartime, but also the specific time of the liturgical year – Advent – when beginning and end, time and eternity, are united. It is not the question, here, to replace philosophy with revelation, but to connect the truth carried by revelation with the living, and vital, experience of every man.

On the one hand, Guardini firmly invites the reader to focus his attention to the concretely living, while, on the other, warning him not to consider faith neither as a mere accessory of the human, nor as something opposed to reason: to raise our gaze towards the spiritual, the symbolic, what lies beyond, allows us to redeem our very rationality by avoiding the dualistic separation between spirit and matter. Only in the living God we find our "homeland". The heart, not the spirit, becomes the driving force. "Heart", here, means a very different force from the sentimentality it began to be associated with only when man's unity was split into the dualism of spirit and matter.

Times is the way man gazes at reality. Faced with eternity and its mystery, human being may take his gaze to the extreme, making an "eternity" out of time's continuous duration, thus trying to make what is finite last infinitely (because beginning and end are unfathomable), or live Eternity as it is, as the cessation of time: eternity is what lasts, while maintaining vitality and generativity. After all, we know and experience a living time, not the empty succession of before-and-after. In fact, depending on our experiences, that same instant can become very long or very short: "It is measured not only by the hands of the clock but also by what is contained within it" (Guardini 1954, 103). Guardini's perspective aims not at denying tradition, but at placing it within a broader scenario: on the one hand, he revives it by affirming the link between being and beings, on the other he is fully contemporary in claiming the temporal character of the events that distinguish us. Man cannot attain *per se* the experience of the pure present of a perfect existence (Guardini 1954, 105), but only by analogy with God:

> If there existed a being whose whole purpose was absolute goodness, a being as powerful as it was good, a being perfectly good, perfectly great, there would no change nor striving in such an existence. The life would be filled with meaning, and its purpose with reality. There

> would be no movement from or towards – the present alone would become absolute. Such a being there is – it is God – and the mode of His being is eternity. Time is not an element that surrounds us, a channel we travel through. We ourselves – our finiteness – are time. Eternity is God's mode of living. (Guardini 1954, 105–106).

Time is neither something surrounding us, nor a channel to cross: instead, it "is" us, our finiteness, just as God, in his infinity, is eternity, because this is the way He lives. How can we become part of it?

> Eternity is not a methaphysical reference that may be expressed through such concepts as "truth", "essence", "life", but that it is a relation between persons. In this lies the special message of the Gospel (Guardini 1954, 108).

It is a place that opens, a space of relationship, a space of beauty. If we perform an action and its fulfillment remains as good, if we lift the frenzy of our days in the stillness of silence and expectation. If we experience the vastness of the natural space around us as parallel and consistent with the inner vastness of our soul, these events are anecdotes that speak of eternity. As always.

Guardini's concreteness is not rationally deduced but allows us to enter an adequate relationship with the concreteness of living. If we can experience eternity as open-ness to relationship, as it is on the theological and properly Trinitarian side, anthropological experience shows us an openness to eternity from experiences of interhuman relationship. What is being played in this open-ness is man's mediating capacity, his ability (to put it in terms of Jewish tradition) to bring redemption closer, or to move it further away if he loses. In the thought of Guardini, the open space of relationship, as the living relationship between time and eternity, can be understood as a communion of opposites that confirms both of them as, at the same time, indispensable and irreducible to each other (Guardini 2016, 42).

3 "The 'Us' Are Eternal". Community, Relationship, and Eternity in Franz Rosenzweig, between Revelation and Redemption (1921)

Moving to Franz Rosenzweig's production, the analogy with the context of Guardini's writing cannot be overlooked. If the Italian-German thinker was already writing at the height of the Second World War, after experiencing the disillusionment for the lack of political education of entire generations of Germans, the Jewish-German thinker moves in the dawn of a new world emerging from the rubble of the 1918, the end of the First World War, but also of a history dominated by the imperial myth. In both cases the present, in its contradictions, is an opportunity to describe a thought overcoming the present itself.

From the critique of the present as an absolute derives and takes shape the sense of an openness to a future made of hope. The commitment of the believer in the world becomes a way of orienting himself to it, knowing that it is not by human action alone that it is guided towards the future.

Coming to properly Rosenzweighian notions, the focus is specifically on the second and the third parts of the *Star*, and, retrospectively, on the first.

The second part shows the fracturing of ordinary time at the hand of revelation's breaking-in, an irruption clearly manifest in the triptych *Genesis-Song of Songs-Psalms*. The creation in *Genesis* (first book of the second part) is the past of the future, the possibility of establishing a future in the possibility of the future: "the creation of the world only needs to find its end in Redemption [*Die Schöpfung der Welt braucht ihr Ende erst zu finden in der Erlösung*]" (Rosenzweig 2004, 130). This first book of the second part of the *Star* denotes a change of style, introducing the dynamic, plastic, and performative language of creation. Only by adequately understanding the theology of creation according to the Jewish-Christian tradition it is possible to fully comprehend the destiny of man and the renewal of creation. Creation is a *chance* for the future. In a recent *Manifesto, Saving Fraternity – Together*, it is emphasized that

> dualism must, once and for all, be amended for the benefit of a new paradigm for life and ecclesial mission, it is what separates – and even creates opposition in – the world of creation (reflected in nature) and that of redemption (outside of nature). This parallelism no longer fulfils the ontological and political function for which it was developed. [. . .] And the freedom of God's creatures, which grants them the honour and the burden of making the world of the living habitable, while waiting for its redemption, is protected by the grace that encourages us to hope for redemption with all our might. This change of tone is crucial for the current *kairós* (Sequeri et al. 2021, 28).

Rosenzweig's perspective anticipates the post-modern drama of the separation between creation and redemption, urging our attention to focus on what is at the same time last and first. Revelation is the ground of the whole building of his thought, as we can see also in the third part of the *Star*:

> Only in the immediate surrounding of that heart of the All, of the Revelation of the divine love, does as much of the Creator and the Redeemer become manifest to us as may become manifest to us; revelation teaches us to trust in the Creator, to await the Redeemer. In this way it lets us recognize even Creator and Redeemer only as the one who loves.

> The one who loves is therefore the only one we immediately behold. As the loving one only God is not the Lord. Here he is active. He is not above his act. He is in it. He is one with it [*Er ist eins mit ihr*]. He loves (Rosenzweig 2004, 404).

Here I give greater importance to the third book of the second part. The redemption of *Psalms* reveals the possibility of living running towards the future, in the fulfillment of time. In the end, the experience of the future that must be accomplished represents a very difficult task, because it requires humility on man's side, at ethical, political, and religious level to avoid believing himself to be the last builder of a definitive construction.

> The world is the other pole towards which the love of neighbor strive. [. . .] The commandment [*Leviticus* 19,18] designates this something as the neighbor [*der Nächste*], and of course, both in the holy language and in Greek, the word means the neighbor as the precise moment of love [*in dem Augenblick des Liebens*] (Rosenzweig 2004, 234).

In our opening up to the others, which opens itself up on the future, we encounter the incompleteness of the world, the impossibility of enclosing the other, the future, history in certain categories:

> God and man already are, the world is becoming. The world is not yet complete [*Die Welt ist noch nicht fertig*]. Laughter and tears are still in it. And the tears are not yet wiped away on all faces. [. . .] If one wanted to give an account of the future, one would surely turn it into a frozen past [*starren Vergangenen*]. That which is future calls for being predicted. The future is experienced only in the waiting [*Die Zukunft wird erlebt nur in der Erwartung*]. Here the last must be the first in thought [*Das Letzte muß hier in Gedanken das Erste sein*]. [. . .] In every inch therefore it is something that is coming, or rather: a coming. It is that which must come [*das was kommen soll*]. It is the Kingdom (Rosenzweig 2004, 235–236).

Not casually, then, the third part [*das Reich zu erbeten*] is in opposition with "*tyrannos*", as the first (about the possibility of knowing the whole of reality [*das All zu erkennen*]) is in opposition with philosophers [*in philosophos!*], and the second part (about the miracle [*das Wunder zu erleben*]) is in opposition with theologians [*in theologos!*]. From this point of view, religious fundamentalists are closely intertwined with political totalitarianisms [*in tyrannos!*] (Caponigro 2015), as well as the soft totalitarianism of consumerist well-being. The incompleteness of the world makes any totalitarian perspective impossible, while, at the same time, removing from the Kingdom the temptation to morph into political messianism (Casper 2003). The anticipation of the Kingdom of Heaven is the idolatrous form *par excellence*. From this confusion, both at theological and political level, arises every form of idolatry in terms of religious and political totalitarianism.

Together we find an authentic idea of the future in the growing Kingdom. "Eternity is not a very long time, but a tomorrow that just as well could be today.

[. . .] Eternity is a today that would be conscious of being more than today" (Rosenzweig 2004, 241).

The moment [*Augenblick*] described in the *Star* is very different from the nihilistic moment of those who live as if there were no future. Surely, this moment is an unrepeatable experience – liturgical, to some extent – of the expected fulfillment. Thus, it is possible to reconcile an oriented reading of history and the experience of the moment as an eternal return (or eternal anticipation) of the future: "the chorus intensifies to the immense unison of the We which pulls with all the united all future eternity into the present 'now' of the moment [*ins gegenwärtige Nun des Augenblicks*]. [. . .] 'But we are eternal [*Die Wir sind ewig*]' " (Rosenzweig 2004, 271). In the eternal "we" is somehow fulfilled the dualistic experience of body and soul: "*In uns eint sich der zweigeborene Strom der Zeit, in unsrer Not, gemeinsamer, dennoch immer geschiedner Not; einfach muss er münden ins ewge Meer*" (Rosenzweig 2003, 63).

Eschatology opens the third book through the events of Israel and the Christian world, both forms of truth differently enacted in history, parallel but not separated, expressions of a truth that is dual but not dualistic, and definitely dialogical. In Jewish life, "past and future, otherwise strangers to each other, the one drawing back when the other's turn comes – here they grow into one [*hier wachsen sie in eins*]: the begetting of the future [*das Erzeugen der Zukunft*] is a direct bearing witness to the past [*Bezeugen der Vergangenheit*]" (Rosenzweig 2004, 317). If the Christian world is the world of time, as the sequence of before and after, the end of Christianity in the secular age is a crossroads [*Kreuzweg*] between the loss of meaning and a new sense of history.

In the third part we find a further element of open-ness to the future: prayer, which establishes the human order of the world, and unites on the Shabbat creation and redemption. It is not a mere cosmic celebration but a creation for redemption. In eschatology, through the conclusions of the *Star*, truth is manifested as dynamic, as an inversion, the return of the Buberian *Alltag des Lebens* which takes up the sanctification of the daily [*Heiligung des Alltags*] of Hasidism. This ultimate truth is also the first for Judaism and Christianity, religions that constantly free themselves from their adherence to the religious, always making the world anew a-religious [*die Welt auch wieder unreligiös zu machen*].

At the same time, this openness manifests the inadequacy of the pre-world: Western thought, in its greatness, hangs in a hypothetical time, a history neither fulfilled nor touched by the event of revelation; the same image of God is necessary to say something about Him, but not sufficient to live the experience of, and with, Him. Even in the present context, we need to replace the hypothetical thought of classic metaphysical tradition not so much with other hypothetical thoughts – even if pragmatic –, such of those techno-science, but rather by giving life to knowledge through the leaven of a way of thinking that is able to dialogue with scientific thought, and to powerfully describe and question our present world with its daring economic, environmental and medical issues.

In the three parts of the *Star* we also find a comparison between Antiquity, when individual life was first superseded by the *polis* and later immersed in the totality of the Roman ecumene, and modernity, when individual life is experienced in the living embrace of the State, generating the dialectics between freedom and political constraint (and, at another level, scientific determinism). While previous epochs strove to build principles in their philosophical, theological and scientific vision, principles that are still valid even if often contradictory, today dialogic thought is called, even in political forms, to speak of these forms of thought in a dialogic, and non-dialectic, perspective: together and alone, freedom and reciprocity, authority and community, uniqueness and individuality. Today the world is definitely a *globus*, even in the pandemic situation. The 21st century is no longer a time of freedom or need, of individual life or social interconnections, but of justice: it is an "open" time. Justice and truth are names of redemption: "Such a simple presence of the Highest One, such an unimpaired all-prevailing and solely prevailing reality, beyond all sorrow and joy of the realization, this is truth. For truth is not, as the Scholastics think, recognized in the error [*am Irrtum*]; truth attests itself [*die Wahrheit bezeugt sich selber*]; it is one with all the real; it does not separate [*sie scheidet nicht*] in it" (Rosenzweig 2004, 406). The way of the one truth is a way against a politics of violence, a politics of division. An erratic attitude could become a way for the truth, an erroneous one becomes a way for division and violence.

4 The Eternity of "I and Thou" in Nishida Kitarō (1932)

The third work we briefly approach here allows us to compare – in a time similar to that in which Rosenzweig and Guardini worked, and only nine years after the homonymous text by Martin Buber – our Western perspective to an Eastern one. In many regards, the Kyōto school deepens and clarifies some typically Western issues.

Only by experiencing an eternal present we can grasp the authentic meaning of the I-Thou relationship. Starting from a perception of time inside and outside man as a fundamental unity of temporality, time is defined as an eternal present (Nishida 1996, 99), not as time's continuum, but as its annulment.

No identity is fixed: in the deep foundations of identity, we can experience the real sense of becoming and temporality (Nishida 1996, 101), as past, present and future (Nishida 1996, 108). Here, in the deep ground of Self – in continuity with mystic Western tradition, for example in Meister Eckhart –, lies the embrace of the I and the Thou, who are found in interiority without ever being reduced to each other (Nishida 1996, 111). The I and the Thou are different, but they meet each other in the ground of soul. Eternity is both the end of time and annulment of time, just as identity is the end of the self and the annulment of the self.

Eastern thinking doesn't see the end of time eschatologically, as the end of history, because this implies a notion of time as moving along a fixed and given direction, favoring instead a circular sense of things – of history and nature. However, this is not without connection with the Christian historical sense of salvation, even if Augustine's philosophy and theology of history redirects history towards fulfillment in its very end. Eternity is revealed as love in practical agency (Nishida 1996, 148–149), in the same way as, in Trinitarian revelation, the love of the Father is revealed in the *kenosis* of the Son through the Holy Ghost.

Finally, in this mystic comprehension of the Self and time, the understanding of the "I" and the "Thou" is always exposed to a radical lack of knowledge, which prevents us from arriving at, and stopping in, a definitive synthesis (Nishitani 1969). In this sense, both Antiquity (Socrates) and the Christian tradition (de Cues) are continuous presences in the Japanese thinkers of the Kyōto school.

The interesting perspective of these Japanese thinkers lies, in summary, in the fact that they offer different ways to understand the connection between time and eternity: phenomenologically, ethically-practically, theoretically, and expressively-aesthetically. At the same time, an Eastern thought that reflects autonomously, while critically confronting Western philosophy, allows us to rethink the original characteristics of the latter. Both the Jewish-Christian path followed by Rosenzweig and Guardini, and the Eastern one of Nishida, are forms of Otherness that force classical Western thought to deeply rethink itself.

5 Conclusion

Therefore, eternity is given not as the overcoming of time in the future, but almost as a priority cessation of time. Thus, the experience of eternity is a daily one (as Guardini's examples show), is liturgical and communitarian (as shown by Rosenzweig's "we"), and deeply rooted in interiority (as the Japanese thought of the Kyōto School demonstrates). In all these perspectives, the intrinsic connectivity of reality makes it possible to speak of the time-eternity nexus both at the epistemological level and as something unveiling within human actions.

References

Caponigro, Gabriella. 2015. *Unde malum? Libertà e tirannia in Franz Rosenzweig*. Pisa: ETS.

Casper, Bernhard. 2003. *Evento e preghiera: per un'ermeneutica dell'accadimento religioso*. Padua: CEDAM.

Casper, Bernhard. 2022. *Il pensiero dialogico. Franz Rosenzweig, Martin Buber e Ferdinand Ebner*. Brescia: Morcelliana.

Guardini, Romano. 1954. *The Last Things concerning Death, Purification after Death, Resurrection, Judgement, and Eternity*. New York, Pantheon Books.

Guardini, Romano. 2016. *L'opposizione polare. Saggio per una filosofia del concreto vivente*. Brescia: Morcelliana.

Nishida, Kitarō. 1996. *L'io e il tu*. Padua: Unipress.

Nishitani, Keiji. 1969. *On the I-Thou Relation in Zen Buddhism*, in "The Eastern Buddhist/New Series", vol. 2, n. 2, pp. 71–87.

Rosenzweig, Franz. 2003. *Il grido*. Brescia: Morcelliana.

Rosenzweig, Franz. 2004. *The Star of Redemption*. Madison (WI): The University of Wisconsin Press.

Sequeri, Pierangelo, Kurt Appel, Carlo Casalone, Dario Cornati, Joao Duque, Isabella Guanzini, Marcello Neri et al. 2021. *Rescuing Fraternity – Together*. Vatican City: Libreria Editrice Vaticana – Pontifical Academy for Life.

Zucal, Silvano. 2004. *Lineamenti di pensiero dialogico*. Brescia: Morcelliana.

Fourth Part: **Existential Corollaries**

Ilaria Malaguti

Eternity, Instant, Duration. *Tangere aeternum*

Abstract: In the intertwining of *chronos* and *Kairos* we can grasp the instant in which, in our temporal and chronological experience, we are offered the possibility of opening ourselves to eternal, *tangere aeternum.* It is possible to recognize in it the call to live the time given to us with full responsibility and to serve the eternal with ardent intensity.

1 To Serve the Eternal? About Idols and Time

In a 1924 text, the Russian philosopher Pavel Florensky wrote: "Men always tend to build idols for themselves to get rid of the hardship of serving the eternal and to passively consecrate themselves to simple givenness" (Florenskij 2020, 98–99).[1]

Immersed in the temporality of our existence, resigned to the uninterrupted flow of days, we accept with resignation that *chronos* is the ruler of our lives. And indeed, Florensky writes, we consecrate ourselves to it, we are destined to a pure givenness that limits our experience to a finite horizon. Perhaps the most singular aspect is that the philosopher recognizes the root of idolatry precisely in consecrating oneself to the pure *factum.* In the wisdom tradition, the idol is the non-living par excellence. The texts insist on the fixity and static nature of idols. "They have mouths, but cannot speak, / eyes, but cannot see. / They have ears, but cannot hear, / noses, but cannot smell. / They have hands, but cannot feel, / feet, but cannot walk, / nor can they utter a sound with their throats" (Psalms 115:5–7). The idol has no breath, no voice, "even though someone cries out to it, it cannot answer; it cannot save them from their troubles" (Isaiah 46:7). Its fixity nullifies the dynamic movement of life. Florensky sees a surprising link between idolatry and time. In its claim to grasp and dominate, in its impatience[2] to have its god at hand, idolatry prevents time from unfolding, cancels it and renounces the *novum* to limit itself to the pure givenness of the fact. The way out of the idolatrous attitude – Florensky

1 Where the English translation has not been indicated, it is understood that the translation is mine.

2 Rabbi Ouaknin writes that idolatry "begins with a gesture of impatience". Impatience is idolatry. See Ouaknin 1999, 42.

https://doi.org/10.1515/9783111313610-020

suggests – lies in the cumbersome service to the eternal. This is a perspective that deserves to be accepted and meditated upon.

In his mythological archetype, Kronos is the god who generates his own children and devours them. In the progressive identification between Kronos and *chronos* (Kerényi 1966), it becomes an expression of man's fragility and vulnerability, a form of the transience of phenomena. Time wears out and consumes. The existence of each man gives reason for the lament that everything passes and nothing remains. This seems to be the law that moves every single existence, tramples it with a supreme and painful ineluctability. And so the risk often arises in us of seeking refuge and consolation in the thought of the eternal, which refers to an indefinite elsewhere, where we can seek reassurance and comfort from the instability and uncertainty of our days. But we risk taking refuge in a thought that alienates us from our existence. Yet what Florensky invites us to do is certainly not seeking refuge in an eternal unconnected with our experience, but to an act of service.

What does it mean to serve the eternal? And is it possible to recognize in the invitation to this taxing service the call to the "responsibility of time"?

2 The Fact and the Instant

Idolatry of time manifests itself in two fundamental ways. First, time is considered and limited to a mere succession of facts. *Factum* is what has been done, completely accomplished, determined and finished. It is an intratemporal phenomenon, it fits into the order of before and after, according to number.[3] The logic of the fact requires that all the stages and passages be followed gradually, according to a precise order, or *ephexes*. The fact is subsumed within a law, it happens in relation to a constituted order, which exemplifies a regularity. This is what Bergson called spatialized time, a time reduced to quantity, to a homogeneity that does not include its nature, but falsifies and empties it.

But what properly defines the idolatry of time is the attitude of people who make the finite instant into the "place" of the infinite and seek infinite satisfaction in it.

The instant has a paradoxical structure. It is not an element of our experience. We do not arrive at it by means of a generalization starting from the data of experience, but by a projection to the limit: the limit of a time without duration. The discourse around the instant seems to lead to an inevitable aporia, that of a

3 See Aristotle, *Physics*, IV, 11.

"timeless temporality". Søren Kierkegaard presented its tragic and painful image in the figure of Don Juan. His search – always dissatisfied – for pleasure empties time, leads it back to a succession of moments devoid of substance, content, deep meaning, and is therefore the source of a dissatisfaction that knows no respite. In Don Juan, the instant becomes a grotesque and idolatrous "parody of the eternal" (Kierkegaard 1980, 86). The infinite succession of instants folds into a movement that constantly returns to itself, in an incessant repetition of pleasure.

But is it possible to think *otherwise* about the instant and its incessant repetition? Can we think of the instant in an amphibolic dimension and glimpse in it the moment in which time and eternity, so to speak, touch each other in our experience? Is it possible to *tangere aeternum*?

In the famous text *The Repetition*, Kierkegaard distinguishes two fundamental types of repetition: one according to nature – proper to natural phenomena and historical-cultural phenomena – the other according to the spirit. This second form or modality is put to the test by experience and in the experience, because it alone can tell if repetition is possible. This is what Constantin Constantius, the protagonist of the Kierkegaardian text, does. Constantin decides to repeat a trip to Berlin, where he had already gone the previous year, "to test the possibility and the meaning of repetition" (Kierkegaard 2009, 20). Once in the German city, he returns to the apartment where he had already stayed, "but no! repetition was not possible here" (Kierkegaard 2009, 21). Everything appears to have changed: the landlord has got married, the room he occupies now "had become dismal" (Kierkegaard 2009, 37), in the cafeteria where he used to go "perhaps the coffee would be just as good as the last time [. . .], yet I did not like it" (Kierkegaard 2009, 37). Constantin then decides to go to the Königsstädter Theater to attend the show *Der Talisman* which, on the occasion of the previous trip, he had admired so much. Now, however, he realizes that the show no longer makes him laugh, that "the little dancer, who had bewitched me the last time with a graceful manoeuvre that resembled the beginning of a leap, had made the leap" (Kierkegaard 2009, 38). The city of Berlin itself now appears, in Constantin's eyes, "prostrate" (Kierkegaard 2009, 22), "the whole city was [. . .] covered in dust" (Kierkegaard 2009, 22). The protagonist therefore experiences that the only thing that repeats itself is the very impossibility of repetition. "I made – he writes – not great discovery, yet it was strange, because I had discovered that there was not such thing as repetition" (Kierkegaard 2009, 38). But at the same time Constantin also discovers the reason for his failure: every attempt at a meticulous reproduction of a past experience had represented *repetition of the wrong kind*.[4] A new way then opens up; this painful failure is not the last word, because

4 See: Kierkegaard 2009, 37.

in the failure of the repetition of the Berlin experience, it is announced that a repetition is possible, even if not in the form of a repetition of the past as a compulsion to repeat. Constantin discovers that what is impossible on the level of an immanent and self-enclosed temporality becomes possible if lived in the dimension of interiority with its openness to the eternal dimension: the possibility of repetition lies in its giving itself as renewal, rebirth, reconciliation.

No longer the return of the identical and much more of a simply unexpected novelty, unexpected or capable of dissolving the ties with the causal series in which it is placed. The authentic and true repetition is qualitatively different. It is a *re*discovered unity, which returns in a renewed form.

Thus emerges the figure who is the true antagonist of Don Juan, the highest model of renewal: Job. Job is the man of sorrows, but – even more intensely – he is the man who experiences the eternal. The three friends who have come to console the great suffering are the expression of a theodicy which proves to be irrelevant. They understand justice as a mechanism internal to the finite, whose function is to re-establish a previous situation. In their presumption, Eliphaz, Bildad and Zophar believe that they can get to the bottom of evil, that they can trace its cause back to an event of the world, invisible to the man of sorrows. This way, not only is the scandal of the pain of the righteous eliminated; even more: the finite becomes the temple of the eternal to which they consecrate themselves. But Job does not see with the eyes of the finite. His word and his gaze go beyond the scene of the world, go beyond the *factum*, reject the repetition of the identical and open up to the eternal. At the end of a long inner struggle, in a dramatic and astonishing dialogue with God, Job bursts into one of the most impassioned prayers that the sacred texts have handed down to us: first – Job says to God – "my ears had heard of you /but now my eyes have seen you" (Job 42:5). Only in the experience of a fragment of eternity is an authentic and very profound inner change possible, only the recovery experienced in eternity can become the *second power of consciousness*.[5] So is repetition possible? Certainly not in immanence, but in a movement that takes up the past and – taking it up again – gives it meaning *again*; that is, it is the root not of a fact, but of an event that renews and therefore never fully accomplished in the finite, never fully present, but always in promise. Repetition is a task for the interiority and is placed in the dimension of the event.

5 See: Kierkegaard 2009, 80.

3 The Event and the Kairos

Sometimes it happens that the orderly plot of chronological time is interrupted; a sudden leap occurs, an unexpected event, which in Greek is expressed by the adverb: *exaiphnes*. It is a term of the Platonic tradition, which occurs in Letter VII, where, dealing with the knowledge of the truth, Plato indicates the need in philosophy for a revolution, an internal change in the soul, similar to the passage from darkness to a day of light; this passage takes place against all expectations, *exaiphnes*, suddenly, at once, "like a light that is lit by the striking of a spark" (Plato, 341C–D). The same term is also found in Dionysius Areopagite, to indicate and express "a process of a thing which, against expectation, is brought to light by conditions that were up to that instant obscure" (Dionysius, 1069 a), something that happens in unexpected way, in relation to which you are not prepared. The event happens independently of us but involves us and demands and determines our response. It is a precise, punctual time, it does not seem connected to what precedes it and what follows it. It is the moment from which a turning point happens, a change of meaning over time. It is the instant that interrupts the linear course of the unfolding of events. The event (*eventum*) is that which takes place and presents something irremediably in excess of its actualization. The excess of latent possibilities is such as to determine a profound change in the one who is affected by it. It is no longer just a matter of happening, but of "making it happen", which offers the possibility of starting, of initiating, of inaugurating; the event is generative of meaning. It completely reconfigures the possibilities of existence. After the event arises, nothing will be as before, the world is no longer the same: the very meaning of the world has changed. Of course, the event is difficult because welcoming it means being willing to reconfigure the world itself, one's being in the world. In this we are called to a very high responsibility and freedom. The event challenges each individual. It breaks in, unhinges and exposes us to an otherness that is beyond all control.

How to walk this path?

In addition to the god Kronos, Greek iconography presents a boy with winged feet.[6] The image is fascinating. His curls are abundant and flowing on the forehead and temples, but sparse in the nape of the neck, so that anyone who tries to grab him from behind inevitably fails. It is Kairos, the decisive and productive moment. Kairos is radically different from Chronos. Évanghélos Moutsopoulos establishes a singular relationship between *chronos*, *kairos* and the different forms

6 It is a bronze statue of Lysippos, made around 330 BC, celebrated by the poet Posidippo.

of dance practiced in the Greek world.[7] Chronos is an expression and model of regularity; he is the dancer of *emmeleia*. "His dance, cyclical, periodic, can be broken down, analyzed, imitated" (Moutsopoulos 1991, 11). With his dance step and his gestures, he "bends the conscience to her rhythmic laws" (Moutsopoulos 1991, 11). Kairos is the surprising acrobat of the *sikinnis*, the dance that "gives its movements uniqueness and totality, challenge and irreparability, risk, commitment and adventure. He invites consciences to remain, like him, intrepid. Chronos the docile, Kairos the indomitable" (Moutsopoulos 1991, 11). *Kairos* breaks the regular rhythm, undermines the inert homogeneity of the flow and embodies the *nyn* of qualitative time, the unrepeatable, the moment that decides meaning. In fact it is "*acmé*, the culminating point, the apogee of life and action, but also [. . .] decisive moment since [. . .] [as we read in Homer] 'everything really lies on the edge of a razor' (Homer, Il., X, 173)" (Zaccaria Ruggiu 1998, 316). Kairos is an expression of the qualitative dimension of time (see Zaccaria Ruggiu 1998, 316). It is not consumed in chronological time, nor does it eliminate it. Rather it installs itself in the *chronos* and becomes, so to speak, the pole that informs the events that precede and follow it, it fits into our existence in a way that breaks and goes beyond the ordered flow of the *chronos*. At a hasty glance, *kairos* may appear analogous to any other event that occurs. But from a spiritual and experiential point of view, it can become a pole around which our entire existence gravitates, it can be chosen as a criterion that directs our existence, illuminates it, substantiates it in its vital fullness and depth, because it is the place par excellence of memory and choice, so that every action that is performed outside of it, or detached from it, becomes impoverished. "Kairos always has a relationship with the divine, whether it is sent directly by the god, or it appears as the 'aionic' point in which the divine breaks in" (Zaccaria Ruggiu 1998, 316). *Kairos* can then become not a space, but a place in which the human being is oriented towards fullness, towards full spiritual realization. In *kairos*, our whole existence, our finiteness, the creatureliness of our words, gestures, affections find their opening, their *plus*. It inaugurates a new time, in which the ordinary fabric of our temporal life is unhinged and, at the same time, an orientation is offered and indicated to our inner life.

Certainly we cannot fail to agree with the ancient adage according to which *factum infectum fieri nequit*. However, in the light of the doctrine of *kairos*, we can understand a passage by Claude Bruaire: "It is necessary to deny – he writes – that the line of time is irreversible" (Bruaire 1974, 72).

7 On the description of the different types of Greek dance, see: Moutsopoulos 1959, in particular see Paragraph 88 and Paragraphs 97–108, dedicated to the dance classification system.

Kairos fits into the fabric of existence, realizes its true reality and unity of meaning, becomes the inner order of our vital experience, the central node of existence. "The entire practical life is thus oriented in a kairic way [*kairique*]; and at the same time, temporality is converted into kairicity [*kairicité*]" (Moutsopoulos 1991, 14). Our life is measured by this: *chronos* does not deliver a long life – in fact we must even recognize that life is short if, however lived for a long number of years, is dispersed in a thousand streams – but the qualitative intensity with which we live the time that is given to us.

How does *kairos* manifest itself? It is easy to think that it breaks in like a whirlwind that upsets and destroys, that it imposes itself with an irresistible force that disrupts the order of things. But the image of the winged boy suggests much more: it suggests movement that urges, impetus that impels, force that presses, but also simplicity, fleetingness, lightness, grace. We can recognize some of the characteristics that define a specific way of being: meekness. "The meek *method*, unlike the remissive or even modest or humble method, knows how to make itself small, but will hardly contemplate its own impotence: it is full power, it is activity in its maximum concentration" (Spinelli 2012, 31–32). Meekness is not inert, it is not submissive; on the contrary: it burns like fire because it exalts action, it is the pinnacle of energy, the culmination of activity, the maximum of subversive capacity, "a concentrated force, reluctant to aggression but not without rebellion" (Spinelli 2012, 20). Can we go as far as to recognize meekness as the mode of manifestation of *kairos*? Certainly, *kairos* is audacious and subversive, it is power that looks into the distance; but it is also patient, it knows how to wait; it waits for time, for *chronos* to be offered, to be recognized, for its action to be effective and transformative and become not only what it sets in motion, but "the very movement of things" (Spinelli 2012, 85).

We often find it hard to recognize *kairos*, we confuse it and align it with the many chronological facts that, *malgré nous*, happen in our existence. But there is no fog to obscure the *kairos* and make it vague; so let's try to make our eye clear, attentive and transparent to the voice of the spirit so that it can recognize the meek event of the *kairoi*.

4 Conclusion

The center of our existence, the actuality of the *ego* with itself, is enclosed in the intertwining of *chronos* and *kairos*. One needs to have time. And we need to give *kairos* time for it to be recognized and welcomed, for it to become incarnate, for it to work in our existence and transform us. Hence its ambiguity: at the same time, *kairos* is and is not part of *chronos*, it does not alienate us from the world, but is irreducible to the logic of the world system.

Can we think of it as the instant in which, in our temporal and chronological experience, we are offered the possibility of *tangere aeternum*? Can we think of the moment starting from an interiority that does not withdraw into itself but becomes attentive and rises in intimate contact with the eternal?

Some passages from the work of Nicholas of Cusa can guide us. To give evidence of the absolute transcendence of God and, at the same time, of his source presence, he suggests the notion of "*Non-aliud*". It is a very profound expression that gives Cusanian thought and language an unprecedented radicality. "Not-other" with respect to opposites is infinity in action as perfect identity. Absolute identity cannot be seen, cannot be known; but it announces itself in the very act of giving rise to the innumerable *identitates*. God is not the totality of entities, but he is totally present in everything, as in the smallest entity. *Non-aliud* therefore says the maximum of transcendence and the maximum of immanence. *Non-aliud* reveals a temporality that opens up to the eternal starting from the point of the most intense interiority lived in experience; and to the eternal it turns not as to a totally other, but as to what it recognizes as its source, root, qualitative origin. From the *Non-aliud* perspective, the analysis of the relationship between time and eternity is reconfigured as a union between a maximum of difference and a maximum of intimacy.

5 *Tangere aeternum*

One of the essential senses of the human spirit is touch. It allows each of us to inhere in what is touched, to experience it, to the point of feeling rooted in it. Touching is an experience of truth. It is not a question of a propositional or judgmental truth but of a truth which is internally experienced; touching opens up a light that brightens because it becomes heat. *Tangere aeternum* does not alienate us from our existence and preserves us from the risk of making the finite into the temple of the infinite. It teaches us to live the time given to us with full responsibility and to serve the eternal with ardent intensity.

References

Bruaire, Claude. 1974. *Le droit de Dieu*. Paris: Aubier Montaigne.

Florenskij, Pavel. 2020. “Note sull’ortodossia”. In *Bellezza e liturgia: Scritti su cristianesimo e cultura*, edited by Pavel Florenksij, 85–101. Milan: SE.

Dionysius, the Areopagite. 1976. *Works*. Translated by John Parker. New York: Richwood Publishing Company.

Kerényi, Kàroly. 1966. *Die Mythologie der Griechen: Die Götter- und Menschheitsgeschichten*. Munich: Deutscher Taschenbuch Verlag.

Kierkegaard, Søren. 1980. “The Concept of Anxiety: A Simple Psychologically Orienting Deliberation on the Dogmatic Issue of Hereditary Sin”. In *Kierkegaard’s Writings*, Vol. 8, translated by Reidar Thomte. Princeton: Princeton University Press.

Kierkegaard, Søren. 2009. *Repetition and Philosophical Crumbs*. Translated by M.G. Piety. New York: Oxford University Press.

Moutsopoulos, Évanghelos. 1959. *La musique dans l’oeuvre de Platon*. Paris: Vrin.

Moutsopoulos, Évanghelos. 1991. *Kairos: La mise et l’enjeu*. Paris: Vrin.

Ouaknin, Marc-Alain. 1999. *Les dix commandements*. Paris: Edition du Seuil.

Plato. 1962. *Epistles*. Translated by Glenn R. Morrow. Indianapolis: Bobbs-Merrill.

Spinelli, Barbara. 2012. *Il soffio del mite: “Beati i miti”*. Magnano: Edizioni Qiqajon.

Zaccaria Ruggiu, Annapaola. 1998. “Aion, Kronos e Kairos. L’immagine del tempo nel mondo greco e romano”. In *Filosofia del tempo*, edited by Luigi Ruggiu, 293–343. Milan: Mondadori.

Santo Di Nuovo

Finitude and Project: For Which Space? And for What Time?

Abstract: After a brief review of the challenges of finitude to philosophies, religions, and sciences, the debate opened by Transhumanism on the possibility of simulating consciousness, and therefore artificially preserving it in time and space, is presented. Heidegger's theme of objective finitude is re-proposed, which leads, from birth to death as a shared universal destiny, to the necessity of a subjective project of life. This realizes the singularity in space and time, but projects itself beyond the time immediately lived, creating a community of projects (not simulated) that challenge the limit of material finitude. Suggestions from Jaspers and Ricœur are discussed. Some conclusions are presented about the ways to implement the project of transcendence, based on Morin's concepts of world citizenship and "reliance".

1 Finitude Challenges Philosophies and Religions

> We seem lovesick for whatever glitters here on earth,
> because we have no knowledge of another way of life.
> What goes on beneath the earth is not revealed to us,
> and we are vainly swept along by fables.
> (Euripides, *Hippolytus*, Verses 195–197)

Euripides' verses tell the relationship between earthly life and an unknown immortal afterlife, which man has desired since the origins of the world. Remaining immersed in the problems of our earthly world, closed in its finite horizon, can we think of the immortal, of infinity, with a limited mind? Can a finite world support an infinite project? Philosophers have been trying for centuries to find answers to these questions.

According to Socrates, because of the immortality of the soul, philosophy should aim to free the soul from the body, by guiding it to the eternal truths, so that, when death comes, it is a liberation from the limits of mortal life. Plato believed that body and soul are two separate entities, the body being mortal and the soul immortal. For Aristotle the intellect is to be immortal. Stoicism attributed immortality only to the universal Logos, conceived as the animating fire of the world, coming from the immanent God and of which individual souls are sparks, which must rejoin it in the final universal conflagration of the cosmic cycle.

https://doi.org/10.1515/9783111313610-021

Augustine and many Christian philosophers resumed the Platonic body-soul dualism. The soul is considered immortal by Scholastic philosophy, albeit with some distinction regarding rational provability, questioned by Thomas Aquinas and William Ockham. Others however (e.g., Averroes) considered the immortal principle not as personal but as impersonal and universal. Much later, Descartes conceived the soul as a substance in itself, whose essential simplicity is the metaphysical foundation of its incorruptibility. For Spinoza, the individual soul collects modes of thought and the extension inherent in the divine substance. Thus the soul participates in the eternity of this substance, as an extra-temporal presence, different from the indefinite extension into future time. Kant considered that demonstrating the immortality of the soul was impossible by rational means, while it can only be "postulated". Otherwise, we cannot conceive the indefinite moral progress, which humanity must propose itself as the main goal of its action, striving to improve in a time indefinite future. Therefore, immortality cannot be known, but must be believed and realized, with an act of faith of "practical reason".

Several contemporary philosophers and psychologists agree that the concept of immortality is unknowable to consciousness. Jung admitted that we are unable to prove that something of us is preserved for eternity, and argued that we can draw on an unconscious part of our mind to have indications about immortality. The problem of immortality is pressing and immediate, and yet deep-seated, and we must dare to make an opinion of it only with the help of clues that come to us from the unconscious. Contrasting the identification of humans almost exclusively with their consciousness, much will still be discovered of what now, with our limited knowledge, seems impossible to us. We live in a world that shaped our minds and established our psychic conditions, so we cannot represent another world, governed by radically different laws of space and time (Jung 1961).

From a sociological point of view, death is not only an individual issue but a social one. Baudrillard (1979) outlines that dead people cease to exist permanently when they are eliminated from the symbolic group representation. Already for ancient Chinese culture, deceased people continued to live as long as someone remembered them and continued their ideas and projects. Bauman (1992) talks about living with death, bidding for immortality, and hypothesizes how modernity can deconstruct mortality, while post-modernity decomposes eternity.

The widespread, even if not conscious, refusal of humans to acknowledge the feared mortality has been underlined (e.g., Ariès 1974; Becker 2011). To overcome the "death taboo" the Death Awareness Movement suggests that people should talk more openly about death, as happens in "Death Cafés" aiming at enhancing human connection about these issues in the "liquid times" described by Bauman (Koksvik and Richards 2021).

Psychosocial theories (e.g., Erikson 1998; Frankl 2000) underlined that the attitude toward death and the afterlife is linked to the process of development in the whole lifespan, and to the meaning of life emerging from it, in close relation with the cultural and social context. Testoni (2003 and 2010) described the relationship of humanity with nihilism, the overcoming of which requires a project that overwhelms the contingency of limited rationality facing these issues.

Going beyond human rational knowledge, several religious faiths have made hypotheses about the mysteries of the afterlife. In some traditions the continuity of life was assured by reincarnation (rebirth of the spiritual principle of an individual in another body) or metempsychosis, transmigration of souls into different bodies, dating back to the Indian Upaniṣad and to Buddhism, and to the Orphic religion, and later reworked by philosophers and writers. But in these conditions, the continuity of individual conscience cannot be assured, since some memories of the previous life could be detected (Stemman 2017). Different forms of persistence after death are believed in other faiths. In Mesopotamian and Greek religions, the Gods made some humans physically immortal. For Christianity, but also for several other religions, all "true believers" will be resurrected to physical immortality in their own bodies. True believers are those that realize in their life the project imposed by their faith. So, both for many philosophies and religions, immortality depends on the actuation of an individual and collective project. Death is not denied, nor trials are made to overcome it, and the efforts are addressed to give it a sense as a door to eternity, to gain by realizing a project of a holy and faithful life.

Not only theologians, philosophers, and writers, but also some psychologists, have cited out-of-body experiences as evidence of a different level of life after death, which overcomes the fear of dying, hypothesizing the possibility of passing to a higher level of consciousness in which what is lost is what humans no longer need: the physical body (Kübler-Ross 1969; Moody 1976).

2 Finitude Challenges Science

Let us consider now the contribution of science to overcome the finitude of human life. According to Harari (2015), for men of science, death is not an inevitable fate, it is simply a technical problem; and every technical problem has its own technical solution. Our best scientists are engaged in investigating every physiological, hormonal and genetic mechanism responsible for disease and aging. They are developing new medicines, revolutionary treatments, and artificial organs that can prolong our lives and, perhaps, defeat death itself.

The search for immortality has a long history already before modern science. Alchemists searched for the stone of eternal life, and in many cultures elixirs of long life were prepared, trying to win the elderly and delay death.

Contemporary biological science has discovered in nature some "immortal" species: Bacteria continuously reproducing through symmetric division (Stewart, Madden, Gregory, and Taddei 2005; West 2003; Ettinger 2005), jellyfishes regenerating through a lineage reprogramming of their cells, so these organisms are potentially open to an indefinite lifespan. Some worms and a genus of medusa called hydra, because of their regenerative capacity, seem to be biologically immortal (Barresi and Scott 2019).

Now technology seems to add unexpected contributions to the solution of the mystery. Technologies supporting life extension aim to reach rejuvenation, i.e. reversing of aging natural process (De Grey and Rae 2007; Rose, Rauser, and Mueller 2011). Cryonics (i.e., low-temperature freezing) supports the hope that a dead body can be revived in the future when medical advancements could avoid death, but this procedure is considered pseudo-scientific. Different sciences (genetics, microbiology, nanotechnology) prospect for the future a "technological immortality" or at least a much longer life. For example, genetic engineering, boosting the amount or proportion of telomerase in the body, makes it possible to double the lifespan of worms. It has been found that mice genetically engineered to produce higher levels of telomerase can prolong their life. Nanorobots inserted within the body could act as an artificial bionic immunity system and delay or invert the aging process.

But a social problem arises, underlined by Harari (2015): Could these technological elixirs of long life be available to all of humanity or only a few rich people who can afford them? This could provoke the anger of those (many) who would be excluded. And in any case, even the lucky ones who enjoy bionic immunity, not being able to avoid being the victim of fatal accidents for different causes external to the individual body, would be left with death anxiety for themselves and their loved ones.

Anyway, immortality or delaying death can hardly be attained by humanity with current scientific advances and in the short future. Modern technologies offer other contributions to the solution of the problem of the mortal finitude of human life.

3 How to Satisfy the Search for Immortality? A Radical Technological Answer

The “immortal life project” of Transhumanism or *Humanity+* (Bostrom 2005) tries to find answers to the finitude based on artificial intelligence. “Augmented humanity” had important precursors in Bacon, Nietzsche, and Julian Huxley. By achieving the fusion of biology, cognitive science, information technology, and nanotechnology, Transhumanism aims at the possibility of blocking the process of deterioration of life. The aim is to create another self, that does not age and does not die. Transcendentalist impulses tend to seek immortality, downloading individual consciousness on the web as a way of ensuring survival after the end of bodily life (Kurzweil 2015).

Experts specialized in neural interfaces aim at giving an ultimate answer to aging pains and diseases and to death itself, transforming frailty-probe bodies and minds into holographic persons (Rosenthal 2018). The ambitious project “Initiative 2045”[1] aspires to replace real, fragile, and mortal persons with potentially immortal holographic people. In the first phase, a realistic copy of a human body remotely controlled has to be created. In the second, an Avatar is programmed in which a copy of the brain is recorded to be transplanted at the end of the personal life. In the third phase, the Avatar will be implemented with the individual personality traits. In the last phase (from 2040 to 2045) a hologram-like Avatar will substitute the dead person for infinite life.

Transforming individuals into realistic avatars would respond to the ancient aspiration of overcoming the physical limits of human life. This would open up the possibility of realizing potential immortality which humanity has aspired to since the origins of the world, that all religions translate into the belief in an afterlife and the possibility of reincarnation, which the simulation could now realize.

Can finitude be defeated by simulation, i.e. creating millions of immortal avatars? It is doubtful if performing a detailed scan of an individual’s original organic brain and simulating the entire structure of his/her personality, can allow reaching the levels of consciousness needed to replicate without time limits the full human existence. The level of detail such scans and simulations would need to achieve to emulate awareness is still to be determined.

1 Itzkov, Dmitry, “Initiative 2045”. http://2045.com/, last accessed 21 September 2023.

4 Searching for a Modern Meaning of Transcendence

To assure some form of overcoming individual death, another way should be followed, i.e. anticipating the construction of something that ensures a "transcendence" within and after the time of life.

The interpretation of transcendence implies some psychological characteristics, i.e. going further into the present structure of the world, thinking differently from stereotyped thoughts, looking for new experiences of humanities, and overcoming the current social and economic limits and constraints. Physical finitude leads from birth to death as universal common destiny. According to Heidegger (1927) birth is "being thrown" in the biological body, in the "finite" world, delimited in space and time. If "Der Mensch ist zum Tode", the awareness of the unavoidable death allows us to open to the "call of care" and to the "authenticity" realized by conscience. During the lifespan, the natural finitude can be faced through a subjective project of life realizing singularity, in space and time. The individual project throws into the immaterial when it connects to a collective project, oriented beyond time individually lived and the finitude of the mortal body. A community of projects can challenge the limits of materiality, expanding in the time dimension, although without a precise chronogram.

Jaspers (1970) affirmed that self-transcendence is a tension towards something that is never reachable, a horizon that is constantly shifting: the law of transcendence is the "ulteriority". This phenomenological approach tries to reconcile two opposing dimensions of life, i.e. the awareness of limitation and the desire to transcend it. Immanence cannot explain human existence, something goes beyond the natural world, which religion grasps but does not exhaust in the human lifespan: the "ways of transcendence". One modality, called *Existenz*, is the open field of possibility unique to every single human being, and constitutes the foundation of the individual self. It is the field of creative spontaneity and freedom. The *Existenz* overcomes natural schemes and constraints and creates new forms of consciousness and spirit (*Dasein*). The second modality of "transcendence", the ground of *Existenz*, goes beyond the natural world of particular objects and configures the "being" itself. The two modes of transcendence in phenomenology can be understood as an opposition to the dehumanizing and objectifying tendencies of modern thought and as the basis of the contrast to an increasingly technological and rationalized culture that loses the substantial dimensions of being-in-the-world (Wahl 1934; Shrag 1971).

5 When Transcendence Fails

What happens when the challenge of transcendence, as proposed by existentialism and phenomenology, is lost in the individual lifespan? The person who is unable to plan or implement the "transcendent" project, and is aware of this inadequacy, experiences existential discomfort which can be also bodily and psychic disease. In Kierkegaard's words, unhappiness is not a condition, but a vocation, i.e. a choice.

The social groups that design projects in their particular egocentric interest, without a "community" perspective, see the prevalence of "domains" such as the economic or technological ones, which no longer remain means but instead ends in themselves. This follows a logic of oppression and supremacy, leading to *Homo homini lupus*. The Latin proverb, cited by Plautus, was retrieved by Hobbes, and shared by Freud who wrote that humans are "creatures among whose instinctual endowments is to be reckoned a powerful share of aggressiveness . . . Homo homini lupus. Who in the face of all his experience of life and of history, will have the courage to dispute this assertion?" (Freud 1961, 58).

But other authors, like as Fromm (1973) state that humans act in the world not only instinctively, but with individual responses to the peculiar circumstances of their life, leading to character orientations based on social and cultural relations. Individualist, egocentric cultures lead humans to transcend nature by destroying others through malignant aggression. Still, they can also create productive relations if the culture and the society support and foster this creative character.

The collectivity that globalizes on a material level, without sharing goals with the "different" people (foreigners, immigrants, disadvantaged . . .), goes towards destruction because it favors being thrown back into finitude, into the common destiny of death. Ethnocentrism and Intolerance produce racism, economic imbalance, famines, conflicts, wars . . . The future of ethnocentric humanity is cultural implosion and social as well as physical entropy: the "black hole" of cultures and societies.

Unfortunately, science (including technology, economy, and finance) can support ethnocentrism and its consequences, expanding "black holes" and fostering negative entropy. A reconnection between science and philosophy is needed to join singularity and universality, existential and cosmological perspectives, as proposed by Heidegger.

6 The Support of Hermeneutics

Philosophy as a hermeneutical activity aims to discover the meaning of existence through the interpretation of phenomena, emergent from being embedded in the world of culture. There is never a symbolic language without hermeneutics; where one man dreams and raves, another man comes forward to interpret (Ricœur 1960). The symbolic is the universal mediation of the spirit between us and the real: the symbolic want to express first of all the non-immediacy of our apprehension of reality.

According to Ricœur, we need to eliminate false and deceptive idols, in the demystifying tradition of the hermeneutics of "suspicion" such as that of Marx, Nietzsche, and Freud. But hermeneutics too aspire to the construction of meaning, allowing to discover symbols and preparing the way for the revelation of significant contents that can totally renew human existence. "Existence becomes a self only by appropriating the meaning, which first resides 'outside', in works, institutions, and cultural movements in which the life of the spirit is justified" (Ricœur, Reagan, and Stewart 1978, 106). Every hermeneutics is, explicitly or implicitly, self-understanding through understanding others, i.e. the meaning of humanity.

The basic meaning of being-in-the-world is that humans are frail and fallible, and evil depends on finitude. The possibility of moral evil is inscribed in the constitution of man. Many "idols" drive humans to interpretations of evil as necessary and unavoidable. But, Ricœur says, we need to destroy the idols to let the symbols speak. We have to decipher the symbols and myths through which the enigma of evil was experienced in Greek and Biblical traditions. Primary symbols (impurity, sin, guilt) are discovered in the myths of "the beginning and the end". The fallibility, if redeemed by the will, is a structure of mediation between the pole of man's finitude and that of infinity (Ricœur 1960).

7 Hermeneutics vs. Sciences

The hermeneutical philosophy, mediating finitude and infinity, can indicate the aims of a project of transcendence. What do science, and technology, offer for supporting this project – very different from the transhumanist project of simulated immortality, quoted above? In the first instance, science can offer the "know-how". From the discoveries of fire and the wheel to antibiotics, to social media, up to artificial intelligence and robotics. But the project cannot be identified with science: Its planning should be personalized and contextualized.

Science is a means, cannot establish the end. "Science does not think", said Heidegger (1993, 374) intending that science "can never reflect on the manner of its proceeding and its tools – does not think, namely, in the manner of the thinker. That science cannot think is not a lack but rather an advantage. That alone secures for it the possibility of engaging in a certain region of objects in the manner of research and settling itself therein" (Heidegger 2000, 133[2]). Science, and technologies, can help in realizing a project but cannot say how the project will be lived by the subjects who realize it.

We can draw an example from projecting houses. No two houses are alike, at least in the interior; construction science tells how a house is best built, but the interior furnishings and above all the affective and emotional atmosphere experienced in the house belong to the owner and his "homing" project. Shifting from *house* to *home*, there are well-built houses, but disordered and unlivable homes, just as there are disordered and unlivable life projects . . . in these cases, we no longer need a construction scientist but a consultant (interior decorator, or other) who helps restore order and livability . . .

The same lack of livable projects is true for psychic disease, social discomfort, uneasiness, and often violence deriving from them. "We're all going to die, all of us, what a circus! That alone should make us love each other, but it doesn't. We are terrorized and flattened by trivialities, we are eaten up by nothing" wrote Charles Bukowski (1998) lamenting the dispersion in the problems of everyday life and the absence of a common project of love, capable of overcoming the common destiny of death. There is a specific aspect of eternity inherent in the bond of love, which overcomes death: this cannot make mortal the promise of eternity containing that bond.

Thus finitude contains in itself the fragment of an unknown feast (Merlau-Ponty 1948): during mortal life, we can see the fragment, not the whole feast. A character from Schnitzler (1896) says that there are perhaps moments that spread a scent of eternity. And the verse of Wisława Szymborska (1996) remember that there is no life that at least for a moment is not immortal.

Isolated sparks, or part of a large fire? The problem is how the individual moments of eternity can be extended and shared in a more general project.

---

2 English translation by Zha (2022), p. 83.

8 How to Implement the Project of Transcendence?

We have to rediscover the uniqueness of the project within the common universal destiny. Neurosciences can participate in interdisciplinary approaches aimed at discovering a general narrative that transcends singularity to achieve collective survival to individual death.

Wilber (2006) proposed the need of shifting from a "small mind" to a "greater mind", extending awareness into transcendental, whatever can be the name attributed to it: e.g., universe, evolution, existence, "the whole", or God; anyway, an anthropological potential deriving by the development of social consciousness. The evolutionary process of mankind can reach high levels of collective intelligence promoting forms of transcendence.

> As human beings we are not only part of evolution – we consciously promote it from within. An essential part of evolution then is the development of higher degrees of intelligence in life . . . [believing] in an in-personal spirit that realizes and recognizes itself in the evolution of intelligent life (Dreitzel 2018, 325).

In a challenging article, Arbib (2016) argues that the soul is a distributed property of our brains and others, and using this shared property some aspects of the individual soul can survive death. Commenting on Arbib's paper, Colagè and Gobbi (2016) present the notion of *self-narrative* as an interdisciplinary concept aimed at capturing essential aspects of persons' identity. Self-narratives are conceived as relational dynamical and distributed constructs, suitable for neural implementations.

> Afterlife can thus be depicted in relational terms as a *unifying system* that embraces and connects all the selves/persons, in such a way that each self has simultaneous and immediate access to everybody else's narrative. Since all the individual selves and narratives converge and relate within such a unifying system, afterlife might be conceived as the ultimate manifestation of the entire history of humanity (Colagè and Gobbi 2016, 258–259).

Some resonances can be found with Stoicism's concept of "universal Logos", which however comes from an immanent God. Also according to John's Gospel, "Logos" is God, but in a transcendent sense. Other similarities can be retrieved with Jungian "archetypes" as universal themes or patterns shared by mankind outside of time. Unlike Jung, the archetypal narrative is not a cumulative knowledge coming from the past, it is instead a shared project to be realized in the future. The question is: what can be the content of the narrative that unifies the individual selves in a common project?

From a different perspective, Edgar Morin (2019) underlines the need for a community of scopes unifying humanity to transcend the present destructive dis-identity triggering poverty, conflicts, wars, and death. The world has to be unified going beyond economic globalization and the political divisions between several realms. A "world citizenship" should be created with the support of education. According to Morin, people should be educated, from childhood, to understand the condition common to all humans, which is grounded on the acceptance of the "necessary and enriching diversity of individuals, peoples, cultures". The awareness of common destiny can save the planet from the destruction caused by the prevalence of the forces of evil (e.g., conflicts between diversities).

Morin created the neologism *"reliance"*, derived from *reli*gion – also in a secular view – and alli*ance.* It is the ability to strengthen positive connections and projects, and/or create new ones. Social identity is possible only if it is connected to our "neighbor", to the community, to the multi-ethnic society, to the human race as a "Hearth-Homeland". This is the "safe haven" for all exiles in search of fraternity, and salvation from the drift to which technology and economy lead, far from aiming at the goal of the common good.

While freedom and equality can be ensured by law, fraternity cannot be enforced. It comes from the recognition of common humanity and respect for its differences. On these bases, the brotherhood between all human beings can be developed to cope with our common destiny of finitude and to share common attempts to overcome it. This road goes towards a project of a common *polis* that world *politics* does not know, or does not want, to orient. The project should concern all areas of human life: economy, work, justice, ecology, health, food, consumption, family, female condition, physical and mental diversity, and aging. All dimensions in which well-being is achieved must find a road that realizes better humanity (Morin 2011). This road seems to recall the Kantian indication about the need for "practical reason" to reach immortality, unknowable by "pure reason"; and Heidegger's proposal of shifting from the individual project to the immaterial collective project, overcoming the finitude of individual lives. It allows reaching *Humanity+* in a different way from Transhumanism. History tells us that many people have been willing to sacrifice their mortal lives to achieve a better life for all following a common project.

9 Last Questions Regarding Faith and Hope

Can we follow the road toward *reliance* using autonomous resources given by the natural endowment, or an external help should be useful? Support, as suggested by several authors, could come from education, sociology, and psychology, and from social health policy. Some additional support could come from faith – religious or not. Faith implies the awareness of interdependence in a common, universal project that transcends time and space and the single person. A project with an outcome unknowable now, for the physical and mental structures in which we are thrown – because it "transcends" them.

Fraternity and love cannot overcome individual death, but can "inscribe a living 'I' in the 'We' of the beloved community . . . the reform of life can only be accompanied by a reform of death" (Morin 2011, 295–296). Love – according to Teilhard de Chardin (1978, 71) – "is the power of producing inter-centric relationship. . . . is the expression and the agent of universal synthesis".

This allows overcoming the question if a *finite* world can support an *infinite*, universal project. The project lives in the hope of a different dimension, of which we cannot foresee the final outcome, but we can see the early fruits already in the present existential dimension. The project of eternity starts in our present world. "It is today that life must become eternal, it is now that we are called to overcome death" (Zundel 1995, 110).

The project of well-being should be widespread and not just reserved for certain social categories (as for technological solutions). The more people feel good, the better is for everyone and for all.

In the housing metaphor reported above, we can start to build more livable and shareable houses rather than shabby or full of goods accumulated without order and without goals. Housing is interdependent, in a condominium built on cooperation and solidarity, which shall survive the individuals living in it.

Following these suggestions, can the finite and entropic world, into which we all are thrown, become a paradise? Certainly not, but it can get as far away from hell as possible. And after? The question remains open . . .

References

Arbib, Michael A. 2016. "Your Soul Is a Distributed Property of the Brains of Yourself and Others". *Reti, saperi, linguaggi: Italian Journal of Cognitive Sciences* 3(1): 5–30.

Ariès, Philippe. 1974. *Western Attitudes toward Death: From the Middle Ages to the Present*. Baltimore: Johns Hopkins University Press.

Barresi, Michael J.F., and Gilbert Scott. 2019. *Developmental Biology* (12th Edition). New York: Oxford University Press.

Baudrillard, Jean. 1979. *L'Échange symbolique et la mort*. Paris: Gallimard.

Bauman, Zygmunt. 1992. *Mortality, Immortality and Other Life Strategies*. Cambridge: Polity Press and Blackwell.

Becker, Ernest. 2011. *The Denial of Death*. New York: MacMillan.

Bostrom, Nick. 2005. "A History of Transhumanist Thought". *Journal of Evolution and Technology* 14: 1–25. http://jetpress.org/volume14/bostrom.html, last accessed 21 September 2023.

Bukowski, Charles. 1998. *The Captain Is Out to Lunch and the Sailors Have Taken over the Ship*. Santa Rosa, CA: Black Sparrow.

Colagè, Ivan, and Nicola Gobbi. 2016. "From Distributed Souls to Relational, Dynamical and Distributed Self-Narratives". *Reti, saperi, linguaggi: Italian Journal of Cognitive Sciences* 3(2): 253–260. DOI: 10.12832/85414.

De Grey, Aubrey, and Michael Rae. 2007. *Ending Aging: The Rejuvenation Breakthroughs That Could Reverse Human Aging in Our Lifetime*. New York: St. Martin's Press.

Dreitzel, Hans Peter. 2018. *The Art of Living and the Joy of Life*. Milan: Angeli.

Erikson, Erik H. 1998. *The Life Cycle Completed: A Review*. New York: Norton.

Ettinger, Robert. 2005. *The Prospect of Immortality*. Ann Arbor: Ria University Press.

Euripides. 2020. *Hippolytus*. Translated by Ian Johnston. http://johnstoniatexts.x10host.com/euripides/hippolytushtml.html, last accessed 21 September 2023.

Frankl, Viktor E. 2000. *Man's Search for Ultimate Meaning*. Cambridge, MA: Perseus.

Freud, Sigmund. 1961. *Civilization and Its Discontents*. New York: Norton.

Fromm, Erich. 1973. *The Anatomy of Human Destructiveness*. New York: Holt, Rinehart and Winston.

Harari, Yuval N. 2015. *Sapiens: A Brief History of Humankind*. New York: Vintage.

Heidegger, Martin. 1927. *Sein und Zeit*. Halle: Niemeyer.

Heidegger, Martin. 1993. *Basic Writings*. San Francisco: HarperOne.

Heidegger, Martin. 2000. *Vorträge und Aufsätze*. Frankfurt am Main: Klostermann.

Jaspers, Karl. 1970. *Chiffren der Transzendenz*. Munich: Piper.

Jung, Carl Gustav. 1961. "Life after Death". In: *Memories, Dreams, Reflections*, edited by Aniela Jaffè. New York: Random House.

Koksvik, Gitte H., and Naomi Richards. 2021. "Death Café, Bauman and Striving for Human Connection in 'Liquid Times'". *Mortality* 16, 380–393. DOI: 10.1080/13576275.2021.1918655.

Kübler-Ross, Elisabeth. 1969. *On Life After Death*. New York: MacMillan.

Kurzweil, Ray. 2015. *The Singularity Is Near*. New York City: Viking Press.

Marcel, Gabriel. 1951. *Le Mystère de l'être*. Paris: Aubier.

Martinez, Daniel E. 1998. "Mortality Patterns Suggest Lack of Senescence in Hydra". *Experimental Gerontology* 33(3): 217–225. DOI: 10.1016/S0531-5565(97)00113-7.

Merleau-Ponty, Maurice. 1948. *Sens et non-sens*. Paris: Nagel.

Moody, Raymond. 1976. *Life after Life*. New York: Bentham.

Morin, Edgar. 2011. *La voie, pour l'avenir de l'humanité*. Paris: Fayard.

Morin, Edgar. 2019. *La Fraternité, pourquoi?* Paris: L'Aube.

Ricœur, Paul. 1960. *Finitude et culpabilité*. Paris: Aubier.

Ricœur, Paul, Charles E. Reagan, and David Stewart. 1978. "Existence and Hermeneutics". In *The Philosophy of Paul Ricœur: An Anthology of His Work*, 97–108. Boston: Beacon Press.

Rose, Michael, Casandra L. Rauser, and Laurence D. Mueller. 2011. *Does Aging Stop?* New York: Oxford University Press.

Rosenthal, Adam R. 2018. "Love of Life: Deconstruction, Biotech & the Survival of Indefinite Life". *Oxford Literary Review* 40(2): 156–180. DOI: 10.3366/olr.2018.0250.

Sandberg, Anders, and Nick Boström. 2008. *Whole Brain Emulation: A Roadmap*. Technical Report. Oxford: Future of Humanity Institute and Oxford University.

Schnitzler, Arthur. 1896. *Liebelei*. Berlin: Fischer.

Schrag, Oswald O. 1971. *Existence, Existenz, and Transcendence. An Introduction to the Philosophy of Karl Jaspers*. Pittsburgh: Duquesne University Press.
Stemman, Roy. 2017. *The Big Book of Reincarnation*. Bophal: Manjul.
Stewart, Eric J., Richard Madden, Paul Gregory, and François Taddei. 2005. “Aging and Death in an Organism That Reproduces by Morphologically Symmetric Division”. *PLoS Biology, Public Library of Science* 3: e45. https://doi.org/10.1371/journal.pbio.0030045.
Szymborska, Wisława. 1996. *View with a Grain of Sand. Selected Poems*. London: Faber & Faber.
Teilhard de Chardin, Pierre. 1978. *Activation of Energy*. New York: Harcourt.
Testoni, Ines. 2003. “L’uomo contemporaneo e il nulla”. In *Visioni Altre*, edited by Remo Bodei, Umberto Curi, Ines Testoni, and Gianni Vattimo, 71–93. Pisa: ETS.
Testoni, Ines. 2010. “Il morire tra ragione e fede”. *Humanitas* 6, 64: 852–905. https://www.research.unipd.it/handle/11577/2418403, last accessed 19 September 2023.
Wahl, Jean. 1934. “Le Problème du choix: L’existence et la transcendance dans la philosophie de Jaspers”. *Revue de Métaphysique et de Morale* 41(3): 405–444. http://www.jstor.org/stable/40897290, last accessed 19 September 2023.
West, Michael. 2003. *The Immortal Cell: One Scientist’s Quest to Solve the Mystery of Human Aging*. New York: Doubleday.
Wilber, Ken. 2006. *Integral Spirituality. A Starting New Role for Religion in the Modern and Postmodern World*. Boston: Integral Books.
Zha, Bowen W. (2022) What Science Cannot Do: The Question Concerning Science and Heidegger. *Open Journal of Philosophy*, 12: 69–85. DOI: 10.4236/ojpp.2022.121005
Zundel, Maurice. 1995. *Vie, mort, résurrection*. Quebec: Sigier.

Diego De Leo

The Last Waltz: Finitude, Loneliness and Exiting from Life

Abstract: Death anxiety and loneliness are major problems for older people. Life expectancy has increased impressively over the last century, but the culture of current societies seems to have great difficulty relating to the limits of finite life. Terror Management Theory teaches us that we tend to imagine that we can reserve the problem of the end of life for others, that is, for those who are "really old". Individuals naturally tend to avoid fears and potentially threatening situations, such as fear of age-related diseases, fear of loneliness in old age and fear of death, situations that stimulate avoidance reactions. When avoidance is not possible, loneliness may become a fierce enemy of the older person. In this case, individuals may find themselves feeling the lack of close attachments, physical touch and intimacy, as well as the deterioration of their health. Depression becomes not only a possible but a frequent evolution, capable of exacerbating the sense of separation from the world and of nourishing self-suppressive ideas. In fact, suicide in old age continues to represent a painfully present eventuality at all latitudes.

1 Old Age Today

Old age, sickness and physical decay seem to be components of life no longer represented in Western societies obsessed with youth and beauty. In a consumerist culture that strongly values youth and physical efficiency, older adults are perceived as a marginal and incongruous presence, perhaps even completely useless.

Among the various conspiracy theories that have circulated about the origin of Covid-19 since the beginning of the pandemic, the systematic elimination of old people – a source of waste of economic resources and unsustainable welfare loads for governments – was certainly among the most hateful. In media, the supporters of this terrible theory reported – as a "proof" of the existing plan – a few lines from a speech held by Christine Lagarde, now the president of the European Central Bank. During her presidency of the International Monetary Fund (2011–2019), while presenting the Global Financial Stability Report 2012, Mrs. Lagarde said, "Old people live too long and this is a risk for the global economy. We must do something, urgently" (De Leo 2020).

However, beyond conspiracy theories, being an older person today – as Bauman argues – is "an incapacity, because it represents a limitation of desires, the

https://doi.org/10.1515/9783111313610-022

moderation of needs, insensitivity to the seductions of marketing: in short . . . it is anathema" (Bauman 2004, 42).

Moreover, if what is called Terror Management Theory (according to which people tend to avoid anything that reminds them of being mortal) constitutes a cognitive defense mechanism necessary to live, it is also true that contemporary society has completely removed the idea of finitude by accentuating the dimension of "fun at all costs" (Valenzano 2016). This appears inserted within a broader strategy of mass distraction with respect to the elaboration of a self-awareness of human existence.

In fact, the need to be "forever young" is leading many old people (not just women) to chase after a lost youth through cosmetic surgery. The sense of limitation, which should characterize advanced age, is not understood but fought against: for example, one dresses exactly like one's grandchildren (wearing uncomfortable tight-fitting jeans and glasses with improbable red frames, for example), instead of adhering to a new self-image, more respectful of the age reached.

After all, today time is increasingly identified as a clock: perfectly suitable for measuring age as the time from birth, but unable to explain age as an indicator of aging processes, and even less suitable for capturing the time lived by being human. Thus, the instrumentalist culture of late modern society has difficulty relating to the elusive but inevitable limits of finite life. This trend is supported by explicit approaches in bio-gerontology that indulge in perspectives of infinite human lives; a message that is eagerly consumed by the mass media and fuels fantasies of longevity that in turn fuel greedy markets for immortality pills. Nutraceuticals with suggestive names ("Elysium", to name an internationally renowned one) acquire authority (and efficacy) by putting a string of Nobel prize winners on their board, knowing too well that older adults are among the first users of vitamins and supplements.

On the other hand, as Valenzano (2016) argues, although life expectancy has increased impressively, this does not seem to have aroused much interest or stimulated significant changes in people's mentality. Modern society continues to have a lot of difficulty in relating to the inevitable end of life. It follows that finitude is easily postponed and reserved only for those who are "really old". The limits of the latter condition are far from shared, but – with excessive optimism – they now seem to tend to consider a century-long life as a legitimate expectation. Indeed, a meaningful understanding of aging should include a process of learning to live a finite life. But we are still far, in my opinion, from this desirable type of education.

Aging is primarily a process of living through changes that are largely beyond our control, although they require an alert and active response. "Unfortunately, aging – the most important experiential source of knowledge about what it means to live a finite life – is neglected by the same culture that so needs its wisdom" (Baars 2017, p 969).

2 Loneliness in Late Life

In the words of John and Stephanie Cacioppo, loneliness describes a negative emotional state experienced when there is a difference between the relationships one would like to have and those that one perceives to have in reality (Cacioppo and Cacioppo 2014). The unpleasant feelings of loneliness are subjective; researchers have found that loneliness is not about the amount of time you spend with other people or alone. Loneliness is more related to the quality of relationships, rather than to quantity. So, "lonely" persons think they do not have significant relationships and, in addition, feel that others do not really understand them.

Studies on loneliness have increased exponentially in recent years. In 2010, 482 articles had among their keywords "loneliness" as listed by Scopus, but in 2021, 2,615 articles had the term "loneliness" among keywords. Lack of social connections has been found to carry a risk of early death similar to physical risk factors such as obesity and smoking (Holt-Lunstad et al. 2015). Loneliness has also been found to be a risk factor for many physical health problems, from fragmented sleep to decreased cardiac output and to dementia (Salinas et al. 2022).

Data from the *Generations and Gender Survey* have clearly indicated that the older adults – especially females – are particularly exposed to the problem of loneliness. The move to retirement living is for many people inextricably connected to their experience of loneliness (Carr and Fang 2021), in addition to the multiple losses usually experienced in late life. These are: reduction in subjective well-being and quality of life; loss of health; loss of physical capacity and functionality; loss in quality of emotional relationships; death of loved ones; reduced social integration; fewer material goods; financial losses; lower cognitive resources; lower perceived mastery; and loss of feeling useful. All these circumstances may feed feelings of loneliness.

"Existential" loneliness has been described as a complex phenomenon and referred to as a condition of life and an experience: when dying and death come closer, existential concerns will come to the fore. A qualitative study by Larsson et al. (2017), based on the narrative interviews of significant others of frail older persons aged 75+ years, showed that the older persons experience existential loneliness when they are increasingly limited in body and space; when they are in a process of disconnecting, and when they are disconnected from the outside world (Larsson et al. 2017).

In another study on older people living in retirement communities across the United Kingdom and Australia, existential loneliness was investigated via qualitative semi-structured interviews. The core themes identified were loss of close attachments, lack of physical touch and intimacy, deterioration of health and body, and lack of an emotional language through which to express own feelings of loneliness as central to older people's experiences (Carr and Fang 2021).

3 Loneliness, Depression and Suicide

Loneliness can lead to feelings of distress, dissatisfaction and detachment (Killeen 1998), as well as silent suffering, fear, anxiety and depression (McInnis and White 2001). Loneliness and depression are related but remain fundamentally different. Loneliness refers specifically to negative feelings about the social world, while depression refers to a more general set of negative feelings (Cacioppo and Cacioppo 2014). Loneliness, isolation, and lack of social interactions are important risk factors for suicide, even after accounting for the influence of mood disorders, which are a common mediator of suicidal behaviors. A modern theory of suicide (Joiner et al. (2009): *The Interpersonal Theory of Suicide*) sees lack of connections (poor sense of belonging) and the perception of being a burden to others as two of the factors necessary for suicidal behavior.

Loss and mourning contribute greatly to the social isolation of those who suffer from it and greatly increase the risk of suicide, especially among men, particularly in old age. A Danish study, based on data from the National Health Register, has shown that the loss of a partner increases the risk of suicide eightfold for older men in the first year after death, a level of risk that continues to remain stable in subsequent years. In bereaved older women, the risk increases five-fold in the first year after the loss but decreases in subsequent years (Erlangsen et al. 2004).

In the Netherlands, there has been much debate on the question whether elderly people over 70 who are tired of life and who consider their life to be completed, should have legal options to ask for assisted dying. In a Dutch study, through 25 in-depth interviews, five main themes emerged: 1) a sense of aching loneliness; 2) the pain of not mattering; 3) the inability to express oneself; 4) multidimensional tiredness; and 5) a sense of aversion towards feared dependence (Van Wijngaarden et al. 2015).

When do old people self-harm? A qualitative study by Wand et al., in Australia (2018) showed that motives denounced by interviewees were taedium vitae ("enough is enough"); hopelessness and endless suffering; cumulative adversity;

untenable situation; being a burden; helplessness with rejection; disintegration of self; and loneliness.

Taedium vitae or weariness or loathing of life in the words of a couple of famous writers, “would cover the famous state of mind, always empirically verifiable, which precedes suicide” (Amery 2012). For Joseph Conrad (2014 [1898]), “Life [*in old age*] is only a desert of lost days . . . There is nothing new, exciting, instructive to expect from the world: no chance to learn something about oneself, gain some wisdom, find fun” (Conrad 2014 [1898]).

“Enough is enough” is a condition that characterized the deliberate exit from life of several world-known personalities. For example, this was the case of Sigmund Feud (Pribor, 1856 – London, 1939), Austrian psychoanalyst; Thomas Szasz (Budapest, 1920 – New York, 2012), Hungarian psychiatrist; Gilles Deleuze (Paris, 1925 – Paris, 1995), French philosopher; Mario Monicelli (Rome, 1915 – Rome, 2010), Italian movie director; Sándor Márai (Košice, 1900 – San Diego, 1989), Hungarian writer; and, Stephan Zweig (Vienna, 1881 – Petropolis, 1942), Austrian writer.

Cases of “enough is enough” are also those represented by the “silent suicides” due to voluntary stopping eating and drinking (VSED). These frequent situations are generally not registered as suicide cases, even if – in an obvious sense – VSED do represent true suicide cases, given that the persons involved truly wanted to die (Wax et al. 2018). Food and water are essential for life: deciding to renounce these elements is quite clearly done to hasten death. So, VSED is a suicide attempt, not just the desire to reduce food and liquids as a natural consequence of some terminal conditions (Quill et al. 2018). VSED has long remained an option to end life in a hidden way; today the debate about VSED is more open and doctors should be well prepared to manage the requests of their terminally ill patients. Since VSED represents a possible response to terminal illness, this contributes greatly to its moral acceptability as a form of suicide (Pope 2017).

4 At Sixty We Can Commit Suicide, at Seventy It Is Advisable, at Eighty Obligatory. Jose’ Bergamin (1895–1983)

This aphorism by the Spanish poet Bergamin suggests one last consideration, which concerns the influence of ageism, i.e., the discrimination based on people’s age that particularly affects older people and further worsens their health and subjective well-being. For older adults, ageism is associated with shorter life spans, poorer physical and mental health, slower recovery from disability, and cognitive decline. Ageism reduces the quality of life of older people, increases their social isolation and loneliness (both associated with serious health problems) and can increase the risk of violence and abuse. The emergency linked to the Covid-19 pandemic has highlighted with brutal clarity the importance of age in clinical care choices.

In its report on ageism (2021), the World Health Organization emphasized that, in addition to the forms of institutional ageism that allow for the rationing of care based on age, there are forms of interpersonal ageism that include ignoring of older people in decision-making processes or avoiding interactions with them; consider them useless because of their age; using an overly accommodating tone and simple sentence structure when interacting with elders (*elderspeaking*). Eventually, older people see themselves as less capable and incompetent, and no longer believe it is possible to learn new skills or take up a new hobby (WHO 2021).

> Rellonen did not give up . . . he was convinced that gathering as a group would certainly have therapeutic effects. It arouses the joy of living and of discovering that things go wrong for others too and that you are not the only unfortunate person in the world (Paasilinna 2006, 129).

References

Amery, Jean. 2012. *Levar la mano su di sé* [*Raising the Hand against Yourself*]. Turin: Bollati Boringhieri.

Baars, Jan. 2017. "Aging: Learning to Live a Finite Life". *The Gerontologist* 57(5): 969–976. https://doi.org/10.1093/geront/gnw089.

Bauman, Zygmunt. 2004. *Work, Consumerism and the New Poor*. London: Open University Press.

Cacioppo, John Terrence, and Stephanie Cacioppo. 2014. "Social Relationships and Health: The Toxic Effects of Perceived Social Isolation". *Social and Personality Psychology Compass* 8(2): 58–72. https://doi.org/10.1111/spc3.12087.

Carr, Sam, and Chao Fang. 2021. "A Gradual Separation from the World: A Qualitative Exploration of Existential Loneliness in Old Age". *Ageing & Society* 43(6): 1436–1456. https://doi.org/10.1017/S0144686X21001252.

Conrad, Joseph. 2014 [1898]. *Heart of Darkness*. Cambridge: Cambridge University Press.

De Leo, Diego. 2020. "Tra congiunti e affetti stabili, andrà tutto bene [Between Family Members and Stable Relationships, Everything Will Be OK]". In *Gli anziani e Covid-19*, edited by Marco Trabucchi, 141–144. Rome: Alpes.

Erlangsen, Annette, Bernard Jeune, Unni Bille-Brahe, and James W Vaupel. 2004. "Loss of Partner and Suicide Risks among Oldest Old: A Population-Based Register Study". *Age and Ageing* 33(4): 378–383. https://doi.org/10.1093/ageing/afh128.

Hansen, Thomas, and Britt Slagsvold. 2016. "Late-Life Loneliness in 11 European Countries: Results from the Generations and Gender Survey". *Social Indicators Research* 129: 445–464. https://psycnet.apa.org/doi/10.1007/s11205-015-1111-6, last accessed 21 September 2023.

Holt-Lunstad, Julianne, Timothy B. Smith, Mark Baker, Tyler Harris, and David Stephenson. 2015. “Loneliness and Social Isolation as Risk Factors for Mortality: A Meta-Analytic Review”. *Perspect Psychol Sci* 10(2): 227–237. https://doi.org/10.1177/1745691614568352.
Joiner, Thomas E. Jr., Kimberly A. Van Orden, Tracy K. Witte, and David M. Rudd. 2009. *The Interpersonal Theory of Suicide: Guidance for Working with Suicidal Clients*. Washington, D.C.: American Psychological Association.
Killeen, Colin. 1998. “Loneliness: An Epidemic in Modern Society”. *Journal of Advanced Nursing* 28(4): 762–770. https://doi.org/10.1046/j.1365-2648.1998.00703.x.
Larsson, Helena, Anna-Karin Edberg, Ingrid Bolmsjo, and Margareta Ramgard. 2018. “Contrasts in Older Persons’ Experiences and Significant Others’ Perceptions of Existential Loneliness”. *Nursing Ethics* 26(6): 1–15. https://doi.org/10.1177/0969733018774828.
McInnis, Gloria J., and Jane H. White. 2001. “A Phenomenological Exploration of Loneliness in the Older Adult”. *Archives of Psychiatric Nursing* 15(3): 128–139. https://doi.org/10.1053/apnu.2001.23751.
Paasilinna, Arto. 2006. [1990]. *Piccoli suicidi tra Amici* [*Hurmaava Joukkoitsemurba* (*Charming Mass-Suicides)*]. Milan: Iperborea.
Pope, Thaddeus M. 2017. “Voluntarily Stopping Eating and Drinking (VSED) to Hasten Death: May Clinicians Legally Support Patients to VSED?” *BMCMedicine* 15(1): 1–3. https://doi.org/10.1186/s12916-017-0951-0.
Quill, Timothy E., Linda Ganzini, Robert D. Truog, and Thaddeus Mason Pope. 2018. “Voluntarily Stopping Eating and Drinking among Patients with Serious Advanced Illness – Clinical, Ethical, and Legal Aspects”. *JAMA Internal Medicine* 178(1): 123–127. DOI: 10.1001/jamainternmed.2017.6307.
Salinas, Joel, Alexa S. Beiser, Jasmeet K. Samra, Adrienne O’Donnell, Charles S. DeCarli, Mitzi M. Gonzales, Hugo J. Aparicio, and Sudha Seshadri. 2022. “Association of Loneliness with 10-Year Dementia Risk and Early Markers of Vulnerability for Neurocognitive Decline”. *Neurology* 98(13). https://doi.org/10.1212/WNL.0000000000200039.
Valenzano, Nicolò. 2016. “L’educazione alla morte tra rimozione, limite e responsabilità [Death Education, between Remotion, Limit and Responsability]”. In *Mors Certa, Hora Incerta, Tradiciones, Representaciones Y Educaciòn Ante La Muerte*, edited by Sara Gonzalez-Gomez, Ivàn Perez-Miranda and Alba Maria Gomez Sanchez. Salamanca: Fahrenhouse Ediciones.
Van Wijngaarden, Els, Carlo Leget, and Anne Goossensen. 2015. “Ready to Give Up on Life: The Lived Experience of Elderly People Who Feel Life Is Completed and No Longer Worth Living”. *Social Science & Medicine* 138: 257–264. https://doi.org/10.1016/j.socscimed.2015.05.015.
Wand, Anne P.F., Carmelle Peisah, Brian Draper, and Henry Brodaty. 2018. “Why Do the Very Old Self-Harm? A Qualitative Study”. *American Journal of Geriatric Psychiatry* 26(8):862–871. https://doi.org/10.1016/j.jagp.2018.03.005.
Wax, Jhon W., Amy W. An, Nicole Kosier, and Timothy E. Quill. 2018. “Voluntary Stopping Eating and Drinking”. *Journal of the American Geriatrics Society* 66(3): 441–445. https://doi.org/10.1111/jgs.15200.
WHO. 2021. *Report on Ageism*. Geneva: World Health Organization.

Luigi Grassi and Harvey M. Chochinov

Beyond the Limits of Mental Illness: Dignity and Dignity Therapy in Person-Centered Psychiatry

Abstract: Amongst the many consequences of psychiatric disorders, such as poor quality of life and poor health, increased physical morbidity, and lower life-expectancy, some are of particularly relevance, such as stigma, discrimination and similar assaults to human dignity. In order to reduce the alienation of those affected by mental illness, person-centered psychiatry is an approach to promote a medicine *of* the person, *for* the person, *by* the person and *with* the person. Dignity-conserving care is part of this approach and it should be practiced in mental health care settings, enabling partnerships with people encountering psychiatric disorders that include mitigating loss of identity, shattering of their self-image, and various challenges within the psychological, interpersonal, spiritual and existential domains. Dignity Therapy, as a personalized and empirically-based intervention developed for patients with life-threatening or limiting illnesses, has been applied in psychiatric settings showing a reduction of stigma and suffering.

> Physicians cannot ignore the many assaults on human dignity, *intrinsic* as well as *attributed*, that are taken for granted in the bureaucratic, commercialized, and impersonal places that hospitals have, all too often, become. [. . .] A more collective sense *of shared responsibility* or the '*dehumanization*,' the '*depersonalization*,' or the '*alienation*' that the sick feel in today's health and medical care institutions must fall on the physician. [. . .]
> (Pellegrino, Merrill, and Schulman 2009, 532; our emphasis)

1 Introduction

According to the World Health Organization (WHO), about 950 million people in the world are affected by psychiatric disorders, of whom almost half have a severe mental illness (SMI) (280 million major depressive, 40 million bipolar, 25 million schizophrenia spectrum disorders), while the rest of the population are affected by less severe disorders that can result in functional impairment interfering with or limiting one or more major life activities (e.g. anxiety disorders alone, about 300 further million people).

Among the many consequences of psychiatric disorders (e.g., poor quality of life poor health, increased physical morbidity lower life-expectancy, risk of sui-

https://doi.org/10.1515/9783111313610-023

cide), stigma, discrimination and violations of human rights are significant issues that are now being broached within the literature (Bhugra, Ventriglio, and Pathare 2016).

For people with SMI, the World Psychiatric Association (WPA) and other scientific associations strongly advocate that having a chronic, incurable disorder, such as certain types of mental disorders (e.g., schizophrenia and bipolar spectrum disorders), should be accepted without guilt or shame, fault or blame. Being seen as an equal human being, with the potential to experience self-worth, meaning and purpose, is a key factor to maintaining dignity, despite suffering the consequences of mental illness. Hence, it is critical to fight for the rights of every person with psychiatric disorders (Burns 2009).

In this chapter we will describe the most significant aspects related to dignity and a dignity-based care in health settings, with specific reference to psychiatry and the need for a person-centered and dignity-conserving approach for people with mental illness.

2 Dignity and Dignity-Conserving Care

Dignity is a core tenet of human life which, according to the dictionary is "the quality or state of being worthy, honored, or esteemed" (Merriam Webster dictionary) or "the importance and value that a person has, that makes other people respect them or makes them respect themselves" (Cambridge dictionary). Dignity derives from the Latin nouns *decus* (ornament, distinction, honor, glory, but also worthiness of honor and esteem) and *dignĭtas (-atis*, derived by *dignus*), which indicate "an individual or group's sense of self-respect and self-worth, physical and psychological integrity and empowerment" (Rosen 2018; Lebech 2011). According to the Kantian perspective, dignity "is an inviolable property of all human beings, which gives the possessor the right never to be treated simply as a means, but always at the same time as an end, because of its ultimate moral worth" (Sensen 2011). In another perspective, dignity can be situated in interpersonal relationships and in the fragile vulnerable finitude of the human individual (Barclay 2016). In medicine and health care systems, dignity is concerned with how people feel, think and behave in relation to the worth or value of themselves and others. To treat someone with dignity is to treat them as being of worth, in a way that it respects them as valued individuals (Jacobson 2007).

For some scholars, dignity relates to the dimension of intrinsic worth that is the inherent and inalienable value that belongs to every human being simply by virtue of being human. The other notion of dignity invokes the dimension of worthiness of respect or attributable dignity, that demands affirmation, and calls for action, approval, and support (Spiegelberg 1971). Sulmasy (Sulmasy 2013) also identifies similar concepts when speaking of *intrinsic dignity*, which refers to worth, stature, or value that human beings have simply because they are human; and *attributed dignity*, thus a created dignity, which refers to worth, stature, or value that human beings confer upon others by acts of affirmation. These two aspects of dignity are sometimes referred to as *dignity-of-self* (the dignity we attach to ourselves as integrated and autonomous persons) and *dignity-in-relation* (the dignity that the individual perceives or does not perceive in the eyes of others within interpersonal relationships).

In medicine there are a myriad of ways in which dignity can be compromised, such as objectification (reification), disregard, indifference, rudeness, condescension, dismissal, intrusion, restriction, labeling, contempt, discrimination, revulsion, deprivation, and assault (Jacobson 2009). Also, the structure of the medical system may favor non-dignified experiences as a result of bureaucratic organizations in which health care professionals are overwhelmed by increasing demands and caseloads, inadequate resources, and uncertainty about the best way to approach their work. Task-based culture prioritizes meeting targets over the provision of patient centered care; spending time filling out forms rather than attending to the psychological and spiritual dimensions of patient suffering; and models of care which value "doing" things rather than "reflecting" on the holistic nature of the patient experience. These are all obstacles to a dignified, compassionate and humanized approach.

On this bases, it is not surprising that patients affected by chronic psychiatric and somatic disorders or life-threatening diseases frequently report a series of interpersonal, psychological and spiritual unmet needs, complaining about the fact that their dignity issues and spiritual pain[1] are not part of the care they receive. For these reasons, undermining of dignity in patients with chronic and disabling conditions causes a sense of meaninglessness of life, loss of identity, and worthlessness of living derived from loss of the future, loss of others, and loss of autonomy. To avoid (or at least reduce) these problems, a dignity approach should instead inform the health care professional-patient relationship by constantly using a framework in which a series of components are intertwined together: compassion, empathy, responsiveness to needs, values and expressed preferences; co-ordination and integration; information, communication and education;

1 Spiritual pain, a notion that aligns with an undermining of dignity, has been defined by Murata as the pain caused by extinction of the being and the meaning of the self, and characterized by a threat to three dimensions of being human: namely a being founded on temporality, a being in relationship, and a being with autonomy (Murata 2003).

physical comfort; emotional support, relieving fear and anxiety; and involvement of family and friends (Cox 2010).

A specific perspective to improve dignity in health care settings has been developed by Chochinov (Chochinov 2002), who suggested that it is possible to structure a dignity-centered approach which can promote a state of physical, mental, social and spiritual wellbeing. According to Chochinov, health care professionals should mindfully explore, in a humanized and compassionate fashion with their patients, specific areas of dignity: the illness-related concerns (i.e., concerns related to symptoms of physical and psychological distress, functional capacity, cognitive acuity) that threaten or impinge on the individual sense of dignity; the issue related to dignity-conserving perspectives and practices for that individual patient (i.e., continuity of the self, role preservation, maintenance of pride, hopefulness, autonomy/control, acceptance, resilience, living in the moment, seeking spiritual help, maintaining normality); and the social aspects of dignity (i.e., privacy boundaries, social support, care tenor, burden to others, aftermath concerns).

3 Dignity-Oriented and Person-Centered Psychiatry

While dignity considerations are important across all medical specialties, (e.g. cardiology, geriatrics, oncology, palliative care), they are particularly salient within the field of psychiatry. The WHO has repeatedly highlighted the importance of promoting principles such as social justice, equality and dignity for people affected by mental disorders on an identical basis with others to end their marginalization and social disenfranchisement (see, e.g., the WHO Mental Health Action Plan 2013–2030, WHO 2021). Respect for the inherent human dignity should also be part of the developing policies, plans and services in the area of mental health. Dignity in mental health refers to an individual's value and worth and is strongly linked to respect, recognition, self-worth and the ability to make choices. The WHO affirms that living a life with dignity stems from the respect of basic human rights, such as freedom from violence and abuse; freedom from discrimination; autonomy and self-determination; inclusion in community life; and participation in policy-making. Amongst mentally ill people, dignity-related issues closely align with considerations of stigma (Livingston and Boyd 2010) and basic human rights (Bhugra, Ventriglio, and Pathare 2016).

Stereotype, prejudice, and discrimination continue to exist, not only for relational encounters in daily life, but within the health care setting and among healthcare providers, with negative consequences on patients' dignity (Mestdagh and Hansen 2014; Corrigan, Watson, and Ottati 2003; Corrigan and Watson 2002).[2] Patients with SMI are ever vigilient to stigma and discrimination, taking various preventive behavioral and psychological measures, such as conscious and strenuous efforts to look, act and behave "normal" (Whitley and Campbell 2014). Severe internalized stigma is related to lower self-esteem, self-efficacy, and likelihood of recovery, whereas severe psychiatric symptoms also have a negative impact on sense of coherence, sense of hope, and quality of life among people affected by SMI, irrespective of the diagnosis (Świtaj et al. 2017).

The literature regarding the relationship between dignity and stigma among people with psychiatric disorders supports the notion that dignity means to be treated as an equal human being, since a lack of compassion in health care setting promotes patient suffering from feeling inferior and stigmatized (Skorpen et al. 2014). Research in psychiatric settings also indicates that dignity is both an intrinsic, self-related process and a reciprocal, extrinsic/interpersonal experience. For instance, an individual's sense of dignity can be thwarted by positive and negative symptoms of SMI, specifically when misunderstood by others. On the other hand, dignity can be enhanced if the patient and significant others embrace a recovery-focused relationship; where they perceive themselves to be treated as individuals, thus reducing shame that may feel associated with being mentally ill (Skorpen, Rehnsfeldt, and Thorsen 2015). Literature indicates that being seen as an equal human being, experiencing dignity despite the disease and frequent marginalization, and fighting for one's own dignity are the main issues expressed by people with SMI. Of course, dignity has been also examined relative to certain psychiatric practices that substantially risk undermining dignity, such as involuntary treatment (Plunkett and Kelly 2021). With respect to this, it can be proposed that stigma and dignity are two sides of the same coin, where self/internalized stigma and social/public stigma correspond, in a negative sense, to what intrinsic dignity (dignity-of-self) and attributed dignity (dignity-in-relation) represent (Fig. 1).

Over the last years, a strong psychiatric movement has developed with the aim of putting the person with mental illness at the center of clinical care and health promotion, at both individual and community levels. More specifically, the Institutional Program on Psychiatry for the Person (IPPP), established by the 2005 General Assembly of the World Psychiatric Association (WPA), started with an initiative af-

---

2 The literature suggests the need to separate the several forms of stigma: institutional stigma refers to the organizational policies or a culture of negative attitudes and beliefs about mental illness; public stigma represents a set of negative attitudes and beliefs that motivate individuals to fear, reject, avoid, and discriminate against people with mental illness; self-stigma is the internalization of public stigma and prejudices influencing an individual's self-conception with secondary feelings of shame, anger, hopelessness, or despair.

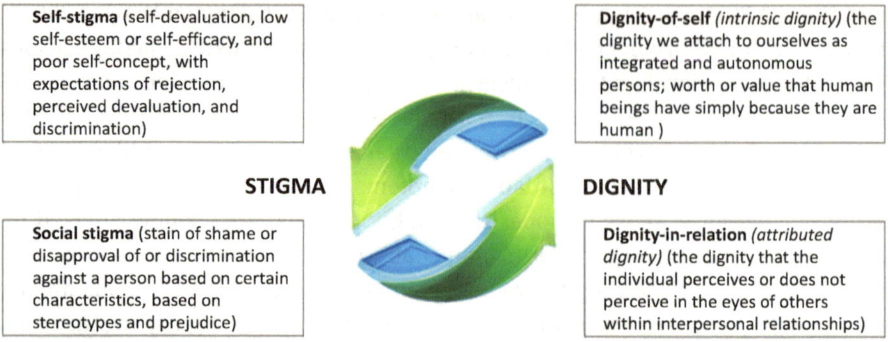

Fig. 1: Stigma and dignity as two sides of the same coin.

firming the need for psychiatry to apply a model based on a whole-person approach. The articulation of science and humanism are the main framework used to optimized all facts of whole persons. Person-centered psychiatry is more than the individualization of care or respect for patients' rights. It also includes recognizing the individual subjectivity of the patient's whole person beyond the limits determined by the illness and its symptoms. This, using the Aristotelian legacy (Cure of a unique person – not of a generalized nosological case – in a specific situation, within a specific, unrepeatable period of one's life), means having health care professionals (as persons) projecting themselves towards the inner world of an individual patient (also as a person) who is in a particular situation marked by suffering and dependence caused by illness. As reported by a number of scholars (Mezzich 2007; Christodoulou, Fulford, and Mezzich 2008), the purpose of person-centered psychiatry is to promote an caring approach *of the person* (i.e., of the totality of the person's health, including its ill and positive aspects), *for the person* (i.e., promoting the fulfillment of the person's life project), *by the person* (i.e., with clinicians extending themselves as full human beings with high ethical standards) and *with the person* (i.e., working respectfully, in a collaborative and empowering manner) (Mezzich et al. 2016).

Person-centered psychiatry starts from the diagnostic process. Rather than "knowledge" (form the Greek διάγνωσις, diágnōsis: διά, diá, "through" plus γιγνώσκω, gignṓskō, "to know") being considered the process by which the nature of a disease or disorder is determined and distinguished from other possible conditions or diseases, knowledge must consider the person first. Such knowledge can be acquired by analyzing all facts that give meaning to that individual person (e.g., history, existential journey, life events, attachment, interpersonal relationships, experiences), allowing clinicians to have a complete description of their health, within the context of their life. Positive aspects of health relate to adaptive functioning, protective factors, quality of life, attending to the totality of the person, including his/her dignity, values, and aspirations. Intervention in person-centered psychiatry should be also organized around the theme of dignity, when using psychotropic drugs, psychotherapy and rehabilitation techniques. In fact, dignity can inform the approach to patients affected by psychiatric disorders, given that healthcare providers have a profound influence on how patients experience illness and that *caring*, besides curing, plays a major role in preserving the dignity of the patients (Chochinov 2013a). As reported by Andorno (Andorno 2013), the patient-health care professional relationship is an opportunity for a careful and active defense of patient's dignity. Studies among patients with acute, mainly psychotic disorders, identified loss of self-identity and anxiety for the future, concerns for social dignity and spiritual life, and loss of personal autonomy as the main factors threatening their sense of dignity.

To help health care professionals routinely explore the dimensions of patients' dignity, the Patient Dignity Inventory (PDI) can be used. The PDI has been developed based on the model of dignity we described (see above) (Chochinov et al. 2008) and, although originally applied in palliative care, its use has subsequently been extended to other medical settings, such as geriatrics, oncology, nephrology, and psychiatry. Factor analysis of the original PDI resulted in a five-factor solution, including symptom distress, existential distress, dependency, peace of mind, and social support. The PDI has been translated into about a dozen languages and validated in several countries to test its role from a cross-cultural perspective. Studies carried out in several countries, including Italy (Ripamonti et al. 2012; Grassi et al. 2017) have shown the PDI to have good psychometric properties.

From the clinical point of view, the PDI can assist clinicians to routinely detect dignity-related distress and guide clinical intervention. In psychiatry, the PDI has been used to help staff elicit dignity-related distress amongst patients with acute psychiatric disorders within inpatient units, helping professionals to improve quality of care and patients to accept treatments (Di Lorenzo et al. 2018). Also, dimensions of dignity were associated with spirituality and psychological well-being among patients with SMI, indicating the need to examine these variables and accordingly implement treatment to achieve recovery-based outcomes (Grassi et al. 2019).

4 Dignity Therapy in People with Mental Illness

Just as person-centered and dignity-conserving care guides the assessment of patients' needs and bio-psycho-socio-spiritual circumstances, so too should it guide the approach to treatment. Among the various and varied interventions (e.g. psychopharmacotherapy, psychosocial rehabilitation intervention) aimed at reducing suffering for people with mental disorders and helping them to regain their function according to a recovery-approach, Dignity Therapy (DT) has a role. DT is based on Chochinov's conceptualization of dignity, and consists of a brief, personalized and empirically-based intervention developed for patients with life-threatening or limiting illnesses. It aims to have patients talk about things that matter most to them, creating a permanent legacy that helps them strengthen dignity and face their suffering. DT aims to enhance the subjective experience of dignity among patients facing conditions that threaten their identity, such as terminal illness (e.g., cancer), chronic life-threating conditions, including more recently diseases that deal with the *sense of the limitation*: limits of memory, such as in dementia; the limits of personhood, such as in psychiatric disorders; or the limit of freedom, such as incarceration (Ounalli et al. 2020; Testoni et al. 2020; Fitchett et al. 2015).

The DT protocol consists in 2–3 individual meeting in which the patient is firstly shown the DT question framework and asked to consider what they might wish to speak about during the session. DT offers the participant an opportunity to reflect upon crucial existential and relational issues, and to review aspects of their lives and of self that they wish to be remembered. This process focuses on tasks such as settling relationships, sharing words of love and wisdom with significant others and preparing a legacy of memories and shared values

About a week later, the DT meeting is carried out and audio-recorded. These meaningful memories, values, words of wisdom, and special messages are transcribed verbatim and then shaped into a narrative through a preliminary editing process. A session is dedicated to the final editing process with the participant, following which they are provided the final written legacy (generativity) document, to be shared and passed along to family members and beloved ones. The ultimate intent of DT is to lessen distress, promote quality of life, validate personhood, alleviate suffering, and give meaning and purpose to life (Tab. 1).

Within mental health, storytelling and narratives have finally been acknowledged as pivotal within recovery-oriented policies (Rennick-Egglestone et al. 2019; Llewellyn-Beardsley et al. 2019). Patients with SMI are confronted with various impairments that derive from their psychiatric condition. Moreover, mental illness may represent a "breakdown" of one's biography, a "profound interruption of the individual's sense of self". Hence, within this specific setting, DT may be adapted to target such existential issues, often overlooked within routine clinical care, by ac-

Tab. 1: Dignity Therapy protocol.

Tell me a little about your life history; particularly the parts that you either remember most or think are the most important? When did you feel most alive?
Are there specific things that you would want your family to know about you, and are there particular things you would want them to remember?
What are the most important roles you have played in life (family roles, vocational roles, community-service roles, etc.)? Why were they so important to you and what do you think you accomplished in those roles?
What are your most important accomplishments, and what do you feel most proud of?
Are there particular things that you feel still need to be said to your loved ones or things that you would want to take the time to say once again?
What are your hopes and dreams for your loved ones?
What have you learned about life that you would want to pass along to others? What advice or words of guidance would you wish to pass along to your (son, daughter, husband, wife, parents, other[s])?
Are there words or perhaps even instructions that you would like to offer your family to help prepare them for the future?
In creating this permanent record, are there other things that you would like included?

knowledging and addressing the ways in which the illness may have altered one's coherent biography, reviewing one's own life, sharing values, thus restoring a sense of personal dignity. In a few experiences of DT in psychiatry settings (i.e., three case reports of patients with depression, alcohol use and schizoaffective disorder; and a randomized clinical trial in patients with major depression), DT appeared to increase sense of hope. In the only DT study intervention available comparing people with SMI (mainly schizophrenia and bipolar disorders) and patients with cancer, similar themes emerged within the generativity documents (Grassi et al. 2022). In fact, qualitative and comparative analysis of DT narratives among both groups of patients yielded similar categories, namely "Meaning" (theme examples "vitality", "self-evaluation", "pride", "evolution of self", "support"), "Resources" (theme examples "support", "resilience", "family", "encounters"), "Legacy" ("bequest for others", "time to say"), and "Dignity", as well as "Stigma" (Tab. 2).

Features of relationships that favor the achievement of optimal therapeutic effectiveness, consistent with a dignity-conserving, person-centered approach include: the creation of a safe space (i.e. an environment or setting in which patients feel safe and secure), clear-cut therapeutic approaches (i.e. skills and techniques used by clinicians to help them communicate and support patients experiencing distress), and a personal growth and self-care disposition i.e. characteristics of the clinician's, which contributes to the therapeutic process, such as the professional's

Tab. 2: Main themes emerging form Dignity Therapy intervention in patients with SMI.

Meaning
"Being with people suffering from a psychiatric condition made me grow. It made me realize that there was a more sensitive world out there, broader than what we might be able to see: a really more important world. Such experience has enriched me" (Marta) *"Being part of the group within the psychiatric rehabilitation center and pursuing that activity, meant finding a meaning to my life, a meaning that I have lost beforehand"* (Fabio).
Resources
"The ability to react positively to obstacles or negative events. Like a tree, if the wind comes, my branches move, but they don't break and I keep my fruit on the branches" (Marta).
Dignity
"Dignity means conserving one's own integrity, being respected and at the same time respecting the others, being treated equally and especially preventing stigmatization" (Giorgio). *"Living with dignity means to have values to express. Everyone has an intrinsic value to express. It only needs to be brought out"* (Marco).
Legacy
"I would like to convey this message: you must always have hope in the future, never let yourself go. Be calm because there is always a way out" (Marianna). *"What is important is that, even if you have failed, you should know that it does not last forever. You can! It's wonderful to have the opportunity and the possibility to say: "I am making it. I want to get back on track and find back the path that I have lost in past"* (Paolo)
Psychiatric illness experience (Stigma)
"[. . .] Sometimes, the others treat me as if I were a sick person (a patient) and this makes me suffer. I say: "I am not (only) a sick person, I am a person who has a few issues but I am not (only) sick" (Valentina) *"My mum is not happy that I am sick. [. . .] She does not easily accept my illness with mental (psychiatric) features. People in general do not accept that and treat us like aliens"* (Fabio)

capacity to maintain emotional health, including self-reflection, openness to personal growth, and acknowledge of one's own feelings of vulnerability (Chochinov 2013b). Other elements include therapeutic humility (e.g. the capacity to not avoid emotion, tolerate clinical ambiguity, explore difficult topics, avoid the urge to have to fix), therapeutic pacing (utilizing approaches that set the clinical tempo and modulate – based on client need – between emotionally containing and evocative strategies, and therapeutic presence (characteristics of the clinician that foster the creation of a safe therapeutic space e.g. being compassionate and empathetic, being respectful and nonjudgmental, being genuine and authentic, valuing intrinsic worth of client).

5 Conclusions

Person-centered and dignity-oriented psychiatry affirm the personhood of the patients and their family members. Within this paradigm, cure and care are complementary, preserving the integrity of the medical profession, including psychiatry (De Valck et al. 2001). Along with communication skills and compassionate care, understanding the multiple dimensions of dignity offers healthcare providers a means to more fully engage their patients, and better attend to their multiple needs and sources of distress.

Some of the burdens and trials of mental disorders can be overcome by paying attention to the salutary effects of human, compassionate interpersonal relationships, including a series of strategies, such as dignity-in-care, that can be applied in psychiatric settings (Chochinov 2022). DT is, amongst the others, an effective method to both help clinicians to better understand their patients, seeing them beyond the limits of their illness, exploring the human side of the person, and helping them to examine parts of their history that they would want known or remembered (legacy and generativity). In our experience, when applied in psychiatric settings, DT led patients with SMI to reveal many dignity-related challenges, including living with stigma, sharing their experience of mental illness and creating a legacy document to help others. DT also helped participants connect with an intrinsic sense of value or worth, with extrinsic reinforcement as a result of sharing the generativity document with those most important to them.

The construct of dignity allows for the reassertion of individual personhood. It should be part of the framework guiding mental health professionals working with people with psychiatric disorders, by acknowledging each person's intrinsic and extrinsic value and going beyond the many limitations mental illness can impose.

References

Andorno, Roberto. 2013. "The Dual Role of Human Dignity in Bioethics". *Medicine, Health Care, and Philosophy* 16(4): 967–973. https://doi.org/10.1007/s11019-011-9373-5.

Barclay, Linda. 2016. "In Sickness and in Dignity: A Philosophical Account of the Meaning of Dignity in Health Care". *International Journal of Nursing Studies* 61 (September): 136–141. https://doi.org/10.1016/j.ijnurstu.2016.06.010.

Bhugra, Dinesh, Antonio Ventriglio, and Soumitra Pathare. 2016. "Freedom and Equality in Dignity and Rights for Persons with Mental Illness". *The Lancet Psychiatry* 3(3): 196–197. https://doi.org/10.1016/S2215-0366(15)00573-8.

Burns, Jonathan Kenneth. 2009. "Mental Health and Inequity: A Human Rights Approach to Inequality, Discrimination, and Mental Disability". *Health and Human Rights* 11(2): 19–31. https://www.hhrjournal.org/2013/08/mental-health-and-inequity-a-human-rights-approach-to-inequality-discrimination-and-mental-disability/, last accessed 19 September 2023.

Chochinov, Harvey Max. 2002. "Dignity-Conserving Care – A New Model for Palliative Care: Helping the Patient Feel Valued". *JAMA* 287(17): 2253–2260. https://doi.org/10.1001/jama.287.17.2253.

Chochinov, Harvey Max. 2013a. "The Secret Is Out: Patients Are People with Feelings That Matter". *Palliative & Supportive Care* 11(4): 287–288. https://doi.org/10.1017/S1478951513000503.

Chochinov, Harvey Max. 2013b. "Dignity in Care: Time to Take Action". *Journal of Pain and Symptom Management* 46(5): 756–759. https://doi.org/10.1016/j.jpainsymman.2013.08.004.

Chochinov, Harvey Max, Thomas Hassard, Susan McClement, Thomas Hack, Linda J. Kristjanson, Mike Harlos, Shane Sinclair, and Alison Murray. 2008. "The Patient Dignity Inventory: A Novel Way of Measuring Dignity-Related Distress in Palliative Care". *Journal of Pain and Symptom Management* 36(6): 559–571. https://doi.org/10.1016/j.jpainsymman.2007.12.018.

Chochinov, Harvey Max, 2022: *Dignity in care*. New Yoork: Oxford University Press

Christodoulou, George, Bill Fulford, and Juan E. Mezzich. 2008. "Psychiatry for the Person and Its Conceptual Bases". *International Psychiatry: Bulletin of the Board of International Affairs of the Royal College of Psychiatrists* 5(1): 1–3. https://doi.org/10.1192/S174936760000535X.

Corrigan, Patrick W., and Amy C. Watson. 2002. "Understanding the Impact of Stigma on People with Mental Illness". *World Psychiatry: Official Journal of the World Psychiatric Association (WPA)* 1(1): 16–20.

Corrigan, Patrick W., Amy C. Watson, and Victor Ottati. 2003. "From Whence Comes Mental Illness Stigma?" *The International Journal of Social Psychiatry* 49(2): 142–157. https://doi.org/10.1177/0020764003049002007.

Cox, John L. 2010. "Medicine of the Person and Personalized Care: A Stitch in Time Saves Nine?" *Journal of Evaluation in Clinical Practice* 16(2): 315–317. https://doi.org/10.1111/j.1365-2753.2010.01391.x.

De Valck, Chris, Jozien Bensing, R. Bruynooghe, and V. Batenburg. 2001. "Cure-Oriented versus Care-Oriented Attitudes in Medicine". *Patient Education and Counseling* 45(2): 119–126. https://doi.org/10.1016/S0738-3991(00)00201-9.

Di Lorenzo, Rosaria, Paola Ferri, Carlotta Biffarella, Giulio Cabri, Eleonora Carretti, Gabriella Pollutri, Ludovica Spattini, Cinzia Del Giovane, and Harvey Max Chochinov. 2018. "Psychometric Properties of the Patient Dignity Inventory in an Acute Psychiatric Ward: An Extension Study of the Preliminary Validation". *Neuropsychiatric Disease and Treatment* 14: 903–913. https://doi.org/10.2147/NDT.S153902.

Fitchett, George, Linda Emanuel, George Handzo, Lara Boyken, and Diana J. Wilkie. 2015. "Care of the Human Spirit and the Role of Dignity Therapy: A Systematic Review of Dignity Therapy Research". *BMC Palliative Care* 14: 8. https://doi.org/10.1186/s12904-015-0007-1.

Grassi, Luigi, Harvey Chochinov, Giuseppe Moretto, and Maria Giulia Nanni. 2019. "Dignity-Conserving Care in Medicine". In *Person Centered Approach to Recovery in Medicine: Insights from Psychosomatic Medicine and Consultation-Liaison Psychiatry*, edited by Luigi Grassi, Michelle B. Riba, and Thomas Wise, 97–115. Cham: Springer International. https://doi.org/10.1007/978-3-319-74736-1_6.

Grassi, Luigi, Anna Costantini, Rosangela Caruso, Serena Brunetti, Paolo Marchetti, Silvana Sabato, and Maria Giulia Nanni. 2017. "Dignity and Psychosocial-Related Variables in Advanced and Nonadvanced Cancer Patients by Using the Patient Dignity Inventory-Italian Version". *Journal of*

Pain and Symptom Management 53(2): 279–287. https://doi.org/10.1016/j.jpainsymman.2016.09.009.

Grassi, Luigi, Maria Giulia Nanni, Rosangela Caruso, Heifa Ounalli, Harvey M. Chochinov, Bruno Biancosino, Ines Testoni, Martino Belvederi Murri, Tatiana Bertelli, Laura Palagini, Silvia De Padova, and Elisa Tiberto. 2022. "A Comparison of Dignity Therapy Narratives among People with Severe Mental Illness and People with Cancer". *Psycho-Oncology* 31(4): 676–679. https://doi.org/10.1002/pon.5913.

Jacobson, Nora. 2007. "Dignity and Health: A Review". *Social Science & Medicine* 64(2): 292–302. https://doi.org/10.1016/j.socscimed.2006.08.039.

Jacobson, Nora. 2009. "Dignity Violation in Health Care". *Qualitative Health Research* 19(11): 1536–1547. https://doi.org/10.1177/1049732309349809.

Lebech, Mette. 2011. *On the Problem of Human Dignity: A Hermeneutical and Phenomenological Investigation*. Würzburg: Königshausen & Neumann.

Livingston, James D., and Jennifer E. Boyd. 2010. "Correlates and Consequences of Internalized Stigma for People Living with Mental Illness: A Systematic Review and Meta-Analysis". *Social Science & Medicine* 71(12): 2150–2161. https://doi.org/10.1016/j.socscimed.2010.09.030.

Llewellyn-Beardsley, Joy, Stefan Rennick-Egglestone, Felicity Callard, Paul Crawford, Marianne Farkas, Ada Hui, David Manley, Rose McGranahan, Kristian Pollock, Amy Ramsay, Knut Tore Sælør, Nicola Wright, and Mike Slade. 2019. "Characteristics of Mental Health Recovery Narratives: Systematic Review and Narrative Synthesis". *PloS One* 14(3): e0214678. https://doi.org/10.1371/journal.pone.0214678.

Mestdagh, Annelien, and Bart Hansen. 2014. "Stigma in Patients with Schizophrenia Receiving Community Mental Health Care: A Review of Qualitative Studies". *Social Psychiatry and Psychiatric Epidemiology* 49(1): 79–87. https://doi.org/10.1007/s00127-013-0729-4.

Mezzich, Juan E. 2007. "Psychiatry for the Person: Articulating Medicine's Science and Humanism". *World Psychiatry: Official Journal of the World Psychiatric Association (WPA)* 6(2): 65–67.

Mezzich, Juan E., Michel Botbol, George N. Christodoulou, C. Robert Cloninger, and Ihsan M. Salloum. 2016. *Person Centered Psychiatry*. Cham: Springer International. https://doi.org/10.1007/978-3-319-39724-5.

Murata, Hisayuki. 2003. "Spiritual Pain and Its Care in Patients with Terminal Cancer: Construction of a Conceptual Framework by Philosophical Approach". *Palliat Support Care* 1(1): 15–21.

Ounalli, Heifa, David Mamo, Ines Testoni, Martino Belvederi Murri, Rosangela Caruso, and Luigi Grassi. 2020. "Improving Dignity of Care in Community-Dwelling Elderly Patients with Cognitive Decline and Their Caregivers. The Role of Dignity Therapy". *Behavioral Sciences* 10(12): 178. https://doi.org/10.3390/bs10120178.

Pellegrino, Edmund D., Thomas W. Merrill, and Adam Schulman. 2009. *Human Dignity and Bioethics*. Indiana: Notre Dame University Press.

Plunkett, Roisin, and Brendan D. Kelly. 2021. "Dignity: The Elephant in the Room in Psychiatric Inpatient Care? A Systematic Review and Thematic Synthesis". *International Journal of Law and Psychiatry* 75: 101672. https://doi.org/10.1016/j.ijlp.2021.101672.

Rennick-Egglestone, Stefan, Kate Morgan, Joy Llewellyn-Beardsley, Amy Ramsay, Rose McGranahan, Steve Gillard, Ada Hui, Fiona Ng, Justine Schneider, Susie Booth, Vanessa Pinfold, Larry Davidson, Donna Franklin, Simon Bradstreet, Simone Arbour, and Mike Slade. 2019. "Mental Health Recovery Narratives and Their Impact on Recipients: Systematic Review and Narrative Synthesis". *Canadian Journal of Psychiatry* 64(10): 669–679. https://doi.org/10.1177/0706743719846108.

Ripamonti, Carla Ida, Loredana Buonaccorso, Alice Maruelli, Elena Bandieri, M. Adelaide Pessi, Stefania Boldini, Caterina Primi, and Guido Miccinesi. 2012. "Patient Dignity Inventory (PDI) Questionnaire: The Validation Study in Italian Patients with Solid and Hematological Cancers on Active Oncological Treatments". *Tumori Journal* 98(4): 491–500. https://doi.org/10.1177/030089161209800415.

Rosen, Michael. 2018. *Dignity: Its History and Meaning*. Cambridge: Harvard University Press.

Sensen, Oliver. 2011. *Kant on Human Dignity. Kant on Human Dignity*. Berlin/Boston: De Gruyter.

Skorpen, Frode, Arne Rehnsfeldt, and Arlene Arstad Thorsen. 2015. "The Significance of Small Things for Dignity in Psychiatric Care". *Nursing Ethics* 22(7): 754–764. https://doi.org/10.1177/0969733014551376.

Skorpen, Frode, Arlene A. Thorsen, Christina Forsberg, and Arne Rehnsfeldt. 2014. "Suffering Related to Dignity among Patients at a Psychiatric Hospital". *Nursing Ethics* 21(2): 148–162. https://doi.org/10.1177/0969733013493216.

Spiegelberg, Herbert. 1971. "Human Dignity: A Challenge to Contemporary Philosophy". *World Futures* 9(1): 39–64. https://doi.org/10.1080/02604027.1971.9971711.

Sulmasy, Daniel P. 2013. "The Varieties of Human Dignity: A Logical and Conceptual Analysis". *Medicine, Health Care and Philosophy* 16(4): 937–944. https://doi.org/10.1007/s11019-012-9400-1.

Świtaj, Piotr, Paweł Grygiel, Anna Chrostek, Izabela Nowak, Jacek Wciórka, and Marta Anczewska. 2017. "The Relationship between Internalized Stigma and Quality of Life among People with Mental Illness: Are Self-Esteem and Sense of Coherence Sequential Mediators?" *Quality of Life Research* 26(9): 2471–2478. https://doi.org/10.1007/s11136-017-1596-3.

Testoni, Ines, Francesca Marrella, Gianmarco Biancalani, Paolo Cottone, Francesca Alemanno, David Mamo, and Luigi Grassi. 2020. "The Value of Dignity in Prison: A Qualitative Study with Life Convicts". *Behavioral Sciences* 10(6): 95. https://doi.org/10.3390/bs10060095.

Whitley, Rob, and Rosalyn Denise Campbell. 2014. "Stigma, Agency and Recovery amongst People with Severe Mental Illness". *Social Science & Medicine* 107 (April): 1–8. https://doi.org/10.1016/j.socscimed.2014.02.010.

WHO. 2021. *WHO Mental Health Action Plan 2013–2030*. www.who.int/publications/i/item/9789240031029, last accessed 28 September 2023.

Ines Testoni

Beyond Alienation: Severino's Removal of Pathological Contradiction

Abstract: Based on a significant reflection by Emanuele Severino on contemporary philosophy (i.e. Severino 1996), this article explores the concept of "alienation" and its evolution from Hegelian theory to psychology/psychiatry and social science studies. The purpose of this article is to highlight the continuity between certain assumptions of Severino's original framework and the comprehensibility of diagnostic aspects inherent to the state of alienation, particularly in cases of schizophrenic spectrum disorders. Severino was a philosopher who presented the most radical understanding of alienation, in line with the concept of nihilism and involving the identification of "being" in relation to its absolute other – that is, "nothingness". The fundamental aim of this essay is to proceed along a path of investigation that considers Severino's thought as the foundation of the epistemology of a new science: one that does not start from the fideistic assumption of the oscillation between nothing and being.

1 Introduction

The concept of alienation underwent a significant decline following the end of the 20th century, until its revival in the context of studies on the centrality of Karl Marx's theories, which cannot be considered outdated because of and despite the Soviet totalitarianism. Rahel Jaeggi's (2014) contribution is of particular importance in this context. She sought to reactivate the discussion around the Hegelian-Marxian key concepts to understand the reasons for the contemporary era's malaise and inability to respond in a nonoppressive manner to the needs of humanity. Her timely phenomenological analysis rekindled interest in this concept but did not resolve the fundamental problem inherent to the reasons why alienation risks to be an obsolete concept. The difficulty consists in the fact that philosophical, epistemological and psychological reflections seem not to be able to exhaustively explain why alienation is an error that can be avoided and corrected. In this article, a proposal for a resolution is presented based on the fundamental contributions of Emanuele Severino.

https://doi.org/10.1515/9783111313610-024

The concept of “alienation”, a term that comes from the Latin word “alienus”, meaning “other”, can be found in the philosophical, sociopolitical and psychological/psychiatric fields; it indicates the estrangement of someone from something or a condition of division of the subject from him/herself. It substantially indicates the separation of subjects and objects that belong together. At its root, this idea refers to a range of cultural, social and/or psychological pathologies involving a self (that can be a person but not necessarily) and an other. Since the separation between a subject and object does not necessarily appear problematic, all related theories focus on making sense of such an estrangement and provide reasons for why this issue is of particular importance, why the separation is challenging, why such a relationship should not be severed and why a reintegration of the divided parts is required. The fields of philosophy, political sciences and psychology/psychiatry consist of different explanations for all these questions, but their intertwined lines of reasoning fundamentally try to intercept the incongruence of separation and determine the difficulties underlying connectedness.

According to Severino (1982; 2015), the nihilism that characterizes the entirety of Western thought and its prephilosophical and mythological precursors forms the essence of the very meaning of alienation. From his point of view, all the explanations of this concept in various fields of knowledge are basically identical and undermined by a pervasive error: nihilism. In the present article, some cornerstones that characterize the discussion on alienation are introduced and connected with the constituent elements of Severino’s originary structure of truth (PST). In particular, they are developed based on considerations of the relationship between psychology and psychiatry, along with the sociopolitical analysis that Severino offered in his work *La filosofia dai Greci al nostro tempo: La filosofia contemporanea* [*Philosophy from the Greeks to our time: Contemporary Philosophy*] (Severino 1996). In the chapter “Scienze umane e decline dell’episteme”, paragraph “Psichiatria e psicologia” [Psychiatry and psychology], the philosopher has stated the following:

> Contemporary psychiatry and psychology [. . .] accentuate their [. . .] relationship with philosophy and the human sciences. And this is a phenomenon that can also be seen [. . .] in all areas of modern science [. . .]. Marxism, phenomenology and existentialism determine to a considerable extent a large part of contemporary psychiatry and psychology. It is an influence that explains the progressive affirmation of the various forms of ‘social psychology’: for example [. . .] the progressive replacement of the medical model, which aims to cure illness by acting on the individual, with the ‘sociological model’ of illness, which aims to establish the extent to which society or the specific social environment in which the individual lives is responsible for his pathological state. In this perspective, the conviction takes hold that the real sick person is not the individual, but the society in which he or she lives, and that the disease of society is capitalist alienation (Ronald Laing, David Cooper, Franco Basaglia).

Thus a true anti-psychiatric attitude takes hold within psychiatric-psychological research, which rejects any technical-therapeutic intervention of the psychiatrist on the patient, based on the conviction that true therapy can only have a political-philosophical meaning, i.e. it is therapy that transforms society (and thus transforms that same psychiatric therapy that in a more or less conscious way is the mouthpiece of social interests and privileges (Severino 1996, 442–443).

Assuming that the psychological and social sciences are increasingly intertwined with and find their critical foundation in philosophy, the aim of the present article is to show the inescapable empirical and heuristic value of Severino's thought in explaining psychological and sociopolitical phenomena, even with respect to the most basic categories of thought concerning alienation. The final objective is to add to previous studies that recognized Severino's thoughts as providing a potential foundation for a new scientific sociopsychological epistemology aimed at liberating people from the anguish of alienation (Testoni 2019; Testoni, Facco, and Perelda 2017; Testoni, Ancona, and Ronconi 2015; Testoni 2021).

2 Alienation of Philosophy and Sociopolitical Thought

The philosophical area focusing on alienation pertains especially to German idealism and the left-Hegelian movement. Within the idealistic system of thought, alienation is an event involved in the development of the spirit that divides and objectifies itself, finally returning to itself in a synthesis. This was particularly the perspective put forward by Georg W.F. Hegel (1807), from whose theory the dialectic concerning the spirit/nature/history's evolution has been derived: the alienation of the spirit ("Entäusserung", "Entfremdung") becomes the other, specifically as nature in space and as history in time. However, *Entäusserung* is also the process of idea formation and then of self-consciousness because the spirit remains itself even when it denies and objectifies itself. *Geist* is the subject that conquers the positive solution through the "determined negation" of alienation ("Aufhebung"), restoring the original form of self-unity. Almost all systematic forms of reflection on alienation are rooted in Hegel's theory.

The left-Hegelian movement immanentized the idea of alienation by applying it to concrete human beings – rather than to the spirit – and their class divisions and history of liberation from ideologies that maintain the social oppression of weaker classes. Feuerbach was the first philosopher to assume the Hegelian concept of alienation and subsequently decline it in a sociopolitical question (Feuerbach 1948). He regarded the Geist as a mere philosophical idea produced by human beings, who constitute the true subject, and then applied the split concept of consciousness to a radical critique of Christianism. As Sigmund Freud (1927) similarly stated in the following century, religion is an instrument of alienation that keeps poorer classes in a state of subordination to those of higher status because it extrapolates the essential characteristics of human beings and attributes them to a fantastic entity, God, who thus becomes the ideal subject for humanity to depend on. According to Feuerbach, this inversion is the essence of alienation, which philosophy aims to show by restoring to humans what authentically pertains to them.

However, according to Feuerbach (1948), Hegel's logic is similar to theology because the spirit corresponds to human thought outside concrete thinking subjects. "To abstract" means "to extract/pull out" or "to alienate". Thus, because of their abstractions, Hegelian philosophy and religion have been founded on the same alienating operation (Feuerbach 1948).

Marx's ideas concerning alienation were greatly influenced by Feuerbachian critical writings. The philosopher took over the immanentization of Hegel's absolute concepts by bringing them back to real subjects, particularly by replacing the parallel idea of the metaphysical concept of state with the idea of a society composed of real human beings who are alienated and desubjectivized by their work. The concept of alienation was then extended to the concrete conditions of the oppressed classes. In *Contribution of Hegel's Critique of Right: Introduction* (Marx 1975 [1843], 175), Marx stated that religion is the "opium of the people". In Christianity, humans in concrete poverty are valued in religious heaven as a sovereign entity. However, the faith of the oppressed in this representation is functional in bourgeois society, wherein formal equality corresponds to real inequality. In contemporary society, individuals are free only in an abstract, nonreal way. The working class is dominated by productive processes, with activities that are mentally and/or physically debilitating, and workers, separated by their products and production processes under the effect of illusion, sell their power to capitalists. This is the alienation of workers that results, on the one hand, in "dispossession/accumulation" dynamics (objects produced being destined for the capitalists who accumulate them and not for the workers) and, on the other hand, in workers' "de-humanization" (consumption of their life and spirit for the benefit of capitalists' richness and well-being) (Marx 1975 [1844]).

Based on the Marxian perspective, the Frankfurt School founded the critical theory of society and developed the idea of dehumanization. The theory's main concept is desubjectification, which is operated not so much and not only by religions but, rather, by the manipulation of individuals through mass media ("culture industry") (Adorno 1966). According to the Frankfurt School, individuals in contemporary culture are reduced to collective standards of behaviors through dynamics of persuasion that transform society into a mass of homologated individuals, functioning towards the needs of mass-production industries. Critical reflection (negative dialectics) thus takes on the task of demystifying mass ideology by shedding light on social onditioning (Adorno 1966). In the diagnosis of Herbert Marcuse presented in *One-Dimensional Man* (1964), individuals in advanced capitalist societies seem to be happy in their conditioned relationships because they identify themselves with their alienating circumstances, thus gaining satisfaction. From his perspective, a small number of individuals in modern consumer societies are empowered to conditionate the perception of freedom by providing masses of individuals with opportunities to buy their well-being and personal happiness. In this state of "unfreedom", consumers behave irrationally by working more than necessary to fulfill real basic needs, ignoring the psychologically destructive effects and the waste and environmental damage it causes (Marcuse 1964).

3 Alienation from a Psychological-Psychiatric Point of View

Psychological and psychiatric studies assume the reduction of any metaphysical dimension to the concrete state of individuals. Then, dissociative phenomenology is produced when the original subject becomes the object, and the objectivized self that is separated from the original self is inherent in patients with a specific biography and bodily health conditions. The aim of psychological and psychiatric discussion is to offer explanatory accounts defining the possible theoretical and empirical solutions that permit the reintegration of the separated parts of patients who cannot behave normally and who have dysfunctional intimate and social relationships (Black and Grant 2014). Thus, mental alienation implies that an individual has become separated from him/herself and the world.

This kind of disturbance is typical of schizophrenic spectrum pathologies. The term "schizophrenia", derived from the Greek words σχίζειν (*schizein*: splitting) and φρήν (*phrēn*: mind), was coined by Eugen Bleuler (1911) and indicates the separation of mental functions. Bleuler (1911) also nosologically introduced a specific feature of schizophrenia, namely ambivalence, which indicates an emotional cognitive state in which contradictory ideas and feelings/emotions are directed towards an object. This is one of the most interesting symptoms on which the incomprehensible behaviors of people suffering from schizophrenic disorders depend. The most important studies that have attempted to decode this symptom and recognize its causes originated in the fields of psychology and psychiatry and were inspired by the contributions of the Frankfurt School. Various concepts of alienation and self-estrangement have been considered to explain internal schizoid states with observable symptoms and external socioeconomic divisions. Political psychology arose from the Frankfurt School and Erich Fromm, with the aim of reversing Adorno's critical dialectic and resolving Marcuse's sociological diagnosis by drawing from the psychological dimension. Studies from this perspective have sought to locate the matrices of alienation in the social dynamics of relationships and communication, revealing how individuals are unconsciously influenced by the forces of power and thus alienated from themselves.

This critical discussion on alienation emerged after a focus on the extreme harms associated with lobotomy, electroconvulsive treatment and insulin shock therapy, and among its ranks were important scholars such as Thomas Szasz, Ronald D. Laing, Franco Basaglia, Silvano Arieti and David Cooper. Others influenced by this wave were Michel Foucault, Gilles Deleuze, Félix Guattari and Erving Goffman, who stand out in importance in the research area inherent to dehumanizing theories that aimed to normalize people suffering from certain forms of alienation. These scholars considered psychiatric treatments to be more damaging than helpful to patients. In line with this position, Umberto Galimberti (1999) in *Psiche e techne: L'uomo nell'età della tecnica* [*Psyche and Techne: Man in the Age of Technique*] highlighted the connection between anxiety and alienation. Specifically, he identified the basis of the alienation and anguish produced by exorbitant amounts of work and stress among contemporary humans as resulting from modern society's demands for indefinite technical development. Human beings living in the most technologically advanced societies are ensnared in an increasingly asphyxiating spiral of production that leaves them unable to give meaning to their daily endeavors. The lack of individual existential purposes leads to alienation and the exponential growth of lifetimes devoted to the production of functional artifacts for the development of global production and technology.

In this regard, Jürgen Habermas (1981) emphasized the role of language and suggested that alienation originates from the distortion of moral debate by dominating market forces and economic power. This kind of distortion results in alienation being extracted from the broader socioeconomic context and the resulting pain and problems being attributed to individual abnormalities or failure to adjust. Habermas's reflections developed under the influence of the antipsychiatry movement and were closely connected to the Frankfurt School's perspective.

The importance of communication and information processing was also the focus of another approach that was strongly influenced by and, in turn, influenced the antipsychiatry and Frankfurt School movements: that of the Palo Alto School, which takes its name from the Californian locality where the Mental Research Institute, a center for research and psychological therapy, is located. Research in this area primarily emphasizes dysfunctional communication or information that conceals substantial contradictions within itself and occurs within vertical relationships. These contradictions are concealed by ambiguity and ambivalence, which are not immediately recognizable and make it impossible for subordinate individuals to understand the authentic thoughts and aims of persons with power over them.

Attention to the problem of dysfunctional communication and information processing was ignited by Gregory Bateson (Bateson et al. 1956). The anthropologist introduced the "double bind theory", according to which contradictions are considered the origins of schizophrenia and post-traumatic stress disorder. A double bind is a dilemma in which an individual (or group) receives two or more contradictory messages. In scenarios involving a differentiation of power or status, this can be dramatically distressing: it generates a situation in which a successful response to one message results in a failed response to the other (and vice versa), so the subordinate person responding will automatically be perceived as being in the wrong, no matter how they respond. Since this contradiction is concealed by ambivalence and ambiguity, the double bind prevents the person from resolving the underlying dilemma or opting out of the situation. This type of communication takes place in a scenario where individuals in power seek to maintain the status quo by keeping their subordinates in a state of disorientation.

At the political level, the most classic example of this kind of contradiction, involving a mix of ambivalence and mystification, is that of Hitler praising peace in *Mein Kampf* (see Testoni 2021). At the psychological level, dysfunctional communication in a family can potentially lead to schizophrenia among the children (Watzlawick, Bavelas, and Jackson 2011). Individuals with mental disorders, which may have originally stemmed from contradictory intimate and social relationships, can experience deep existential alienation within their communities due to other people's, and potentially their own, negative attitudes towards them and result in dysfunctional behavior involving continuous attempts to adapt to an essentially hostile environment. This wide area of research on communication contradictions has revealed that, in today's consumerist society, individuals are estranged from their sense of self due to the repressive injunction to be happy; such an injunction does not allow room for the recognition of alienation and could be seen as an expression of alienation itself (Habermas 1981).

As Severino (1996) indicated in his reading of contemporary philosophy, the concept of alienation, interesting as it is, runs the risk of being insubstantial since relations are no longer considered necessary. With the decline of metaphysics – particularly the aspect of metaphysics that Severino defined as "epistemic" and was aimed at demonstrating the necessity of connections between the parts of a whole or totality – contemporary epistemology considered all relations to be merely probable and thus fundamentally linked to chance and entropy (i.e. Prigogine 1978); in addition, scientific knowledge has been historically determined as fallible (i.e. Popper 1959). On the basis of this assumption, which came to light in its entirety with the most radical coherence of the idea of contingency developed philosophically by Nietzsche and Leopardi (Severino 1996, 1999) and assumed by the hard sciences, no integral unity of any subject is necessary.

However, thanks to Severino's indication, the theme of alienation once again gained importance and became valuable in the psychological-psychiatric sphere to understand common ways of reasoning as well as the extreme sense of suffering that nihilism entails and that manifests in mental disorders. In fact, the solution to the concept of alienation can be obtained from an awareness of the substantial dynamics arising from nihilist contradictions.

4 The Substantial Madness of Western Thought and Its Solution

Ronald D. Laing was an influential psychiatrist in the area of psychological research that emphasized the role of contradictions in schizophrenia. In line with the antipsychiatry movement and the Frankfurt School and Palo Alto School's course of research, the psychiatrist argued that problematic families and socio-economic oppression can cause alienation or "ontological insecurity" in individuals (Laing 1967). Going against how mainstream psychiatry and society diagnosed alienation disorders, he stated that schizophrenic symptoms could be considered adaptations to dysfunctional environments (Laing 1967).

Emanuele Severino, in his work *Techne: Le radici della violenza* [*Techne. The Roots of Violence*], paragraph "Il segreto della follia" ["The secret of madness"], quoted Ronald Laing (Severino 1979b, 283–284), as given below. In the lines preceding this quote, Severino reported what Laing (1959), in *The Divided Self. Study of Existential Psychiatry*, wrote about one of his patients with schizophrenia, who jokingly began to perceive the dissolution of her own identity, which then merged with the trees in a forest to the point that she could no longer return to herself.

> The game consists in pretending to "disappear", to leave "empty" the place one occupies on earth, "blending in" with what is around; it consists in pretending no longer to be there ("and I would no longer be there"). Then the game becomes serious: the little girl in the park ends up convincing herself that she is "blending in" with the world around her and "no longer being there". But when this happens to her, anguish over the nothingness she sees herself becoming also sets in. "Then I call myself by name many times, as if to make myself come back". The little girl in the park becomes an irredeemable psychotic when, despite every effort, she can no longer "make herself come back" into existence and remains

> there in nothingness. [. . .] This is not an isolated case. Psychological and psychiatric analyses record a large number of cases in which the sick person is convinced of "not being there any more", like the little girl in the park, or of never having been (since childhood, his parents "treated him as if he were not there"), or of not being himself, but another person, or a thing – stone, fire, wood, ornament. Even in cases where the sick person identifies himself with something else (person or object), he has reduced himself to a nothingness.

Indeed, many schizophrenic spectrum disorders present similar symptoms. For example, the so-called "Cotard's Syndrome" is characterized by an inability to perceive oneself, and patients interpret their emotions as being dead. In such cases, anguish is expressed along with anhedonia, which corresponds to the feeling of annihilation. The Capgras delusion presents the conviction that a close individual (a friend, family member or pet) has been replaced by an identical impostor. For both these expressions of psychological imbalance, the difficulty lies in clearly defining why these persons are "alienated".

Severino considered the schizophrenic woman described by Laing a prototypical example of the madness of all Westerners, as they had inhabited the nonsensical perspective of nihilism. As the philosopher indicated, at the basis of the possibility of understanding any statement, there is a reference to the abysmal difference between truth and error (Severino 1982; 2015). Madness is the negation of truth, and alienation is the expression of an error. Based on *La struttura originaria* [*The Originary Structure*] (1981) and *Essenza del nichilismo* (Severino 1982, *The Essence of Nihilism*, Severino 2015), it is possible to recognize the very first axes that permit one to understand the essence of alienation as an error that can only be recognized by the truth. The concept of truth proposed by Severino refers precisely to the incontrovertible, the structure of which shows the self-contradicting and self-negating nature of statements that want to deny the true assertion. The PST and, more precisely, the "originary structure of the destiny of truth" are based on the "primary foundation" of the eternity of being, which involves four axes on which the identity of being with it/oneself and not with nothing is pivoted.

In the first axis, the foundation of the originary structure of truth over arguments that seek to negate truth occurs through the *élenchos* – the argumentative dynamic that shows the self-negation of erroneous (folly) content. *Élenchos* is based on the undeniable fundamental opposition of the positive (affirmation) and negative (negation). This first axis is related to the second axis, which involves the principle of noncontradiction (PNC; or the law of contradiction) (Aristotle): for all propositions p, it is impossible for both p and not p to be true; symbolically "~(p · ~p)", in which ~ means "not" and "·" means "and". In the chapter "Returning to Parmenides" in *The Essence of Nihilism* (Severino 1982; 2015), Severino reiterated the incontrovertible assertion that being is not and cannot be nothingness and that nothingness is not and cannot be being. In discussing Aristotle's *Meta-*

physics IV, Gamma, 3–6, he affirmed that without the PNC, we cannot know anything that we do know. In other words, if I affirm that being is being, I am not affirming that being is not being, or that being is nothing.

The first two axes imply the following axial elements:

- The third axis presents the principle of identity: A ≡ A; every being is identical to itself, or "(∀x) (x=x)", in which ∀ means "for every" or simply that "A is A". Another formulation of the principle, derived from the basic form just defined, asserts that if a propositional function F is true of an individual variable x, then F is indeed true of x; symbolically F(x)   F(x), in which   means "formally implies". In sum, being is being.
- The fourth axis involves the principle of the excluded middle (or third) (PEM): either p or ~p must be true, with no third or middle true proposition between them; symbolically, "p ∨ ~p", in which "∨"means "or". This implies that, between affirmation and negation, there is no third proposition.

Then, eternity means that it is necessary that each being be and be as it is (A ≡ A). It is impossible for any being not to be (~[p · ~p]). Everything that appears is not nothing but a being and thus eternally itself. Everything that appears is and is forever, since whatever is cannot come into being from nothingness or cease to be by falling into nothingness. "Appearing" means entering into the horizon of experience. Nihilistic alienation is the identification of a being to its negation – that is, to nothingness. Since nothing is not being, beings cannot turn into nothingness, and nothing cannot really turn into beings. Any faith in considering beings as oscillating between nothingness and being involves the alienation of being.

The second, third and fourth axes of the TSP can be applied to the concept of psychic alienation in a similar way. Severe psychological/psychiatric disorders begin to be recognized when a person no longer identifies him/herself and others, when s/he confuses him/herself and others with something else (e.g. "I am not myself; I am a tree." "My mother is a chair.") or when s/he is unable to differentiate affirmation and negation (e.g. "Maria is my mother; Maria is not my mother." "Antonio is my brother; Antonio is not my brother.") and keep them separate (e.g. "I do not know whether Maria is my mother." "I do not know whether Antonio is my brother."). These three aspects are all reducible to basic conceptions: "I do not exist"; "a part of me does not exist"; "my mother Maria does not exist"; "my brother Antonio does not exist".

Alienation is the clearest expression of betrayal. Indeed, Severino has shown how each of the founding traits of the concept of truth belongs to the history of Western thought as well as how Western thought has betrayed its own intentions and failed to manage these principles. Westerners want to truly understand the sense of being and of reality but, to the contrary, are unable to respect all the fundamental axes of truth. The basis of betrayal lies in the claim that the appearance of "becoming" in the world amounts to the appearance of the annihilation of beings that become. Nihilistic alienation that characterizes the entire history of Western thought considers beings to be more or less destined for annihilation. The conviction that becoming, seen as inherent to contingent beings, exists in the oscillation between being and nothing assumes that there is a time when being is nothing; it *simpliciter* assumes that A is not A, affirming and negating that A is A.

Faith in becoming – that is, the belief that beings that inhabit experience oscillate between being and nothingness – is a fundamental error. The impossibility of the existence of a time in which beings that appear are "not yet" or "no longer" is the foundation of the impossibility of creation and annihilation; "it is impossible" means that the claim that "a being is not" negates the PST structure. More specifically, it is a negation of the opposition between the positive and the negative and would mean affirming that a being is "other" than itself. Alienation consists of the negation of the fundamental principles of any logical assertion: the principles of identity, noncontradiction and the third excluded. This implicit negation is the basis of the auto-negation that the *élenchos* of the PST explicitly describe.

Further, another axis can be added, one that involves the scientific observation of phenomena and it is inherent to the way in which humans observe and interpret facts. In *Legge e caso* (Severino 1979a; *Law and Chance*, Severino 2023) Severino added an "Introduction" to his translation of Carnap's *Der Logische Aufbau der Welt* and discussion of Carnap's philosophy. The strict analysis of the logical positivist position that is inherent in what appears was a significant task, which Severino undertook precisely to analyse one of the most coherent manifestations of contemporary thought. This rigorous analysis shows that while Severino's concept of "appearing" was radical, it was also so cogent that it could be assumed by the most thorough neopositivist thinkers. In fact, the most severe observation of everything appears indicates that it is impossible to say that one can observe an act of creation and an act of annihilation. It is absolutely impossible to say that "nothingness" appears.

5 Conclusion

What Marxism, the Frankfurt School, antipsychiatry and the Palo Alto School succeed in clueing in is that a process of social alienation underlies the madness of individuals, which is based on an erroneous faith that mystifies the meaning of what it really is. However, their analysis fails to authentically explain what this erroneous faith is based on – what the most authentic and radical form of alienation consists of – which Hegel himself failed to identify. For this reason, the concept of alienation has declined substantially in recent years. Today, however, it is possible to resume this path of reflection and free the inhabitants of the West, who believe they are annihilable and, therefore, that their being can be nothing from the oppression of anxiety.

Severino has shown that, contrary to Western philosophic assumptions, no becoming appears in the sense of the appearance of the annihilation or the becoming ex nihilo of beings. Therefore, it is incorrect to say that Severino's perspective denies experience and that claiming the eternity of every being amounts to denying the manifold display of appearance. On the contrary, Severino contended that the content that actually appears in no way testifies to the annihilation and creation of beings. Believing that the scenario in which we are immersed testifies that creation and annihilation are hallucinated by alienating faith. To consider the "becoming" testified by experience as the coming from or returning to nothingness is thus the content of madness.

Furthermore, the philosopher's theory affirms that humans suffer because of what they believe they are; if they believe they are mortal (i.e. annihilable), their suffering can only be extreme and incurable. The anguish of the contemporary period is due to the fact that humans think that they are totally annihilable or are convinced that they are nothing. Since this thought is at the basis of any reasoning, everything loses sense, and the anguish that arises is terrifying. However, appearance and experience cannot attest to the fact that what no longer belongs or what does not yet belong to experience has become nothing or remains nothing. Thus, it must be the case that every variation in the world is the beginning of the appearance of eternal beings.

Knowing all this is not the same as not knowing it, simply because of the undeniable and logical understanding of the concept of "eternity".

Feeling alienated due to the anguish and fear of being nothing? What about eternity?

References

Adorno, Theodor W. 1966. *Negative Dialektik* [*Negative Dialectics*]. Berlin: Suhrkamp.

Bateson, Gregorio, Don D. Jackson, Jay Haley, and John Wealland. 1956. "Toward a Theory of Schizophrenia". *Behavioral Science* 1(4): 251–264. https://doi.org/10.1002/bs.3830010402.

Black, Donald, and Jon E. Grant. 2014. *DSM-5 Guidebook: The Essential Companion to the Diagnostic and Statistical Manual of Mental Disorders* (5th edition). New York: American Psychiatric Pub.

Bleuler, Eugen. 1911. *Dementia praecox oder Gruppe der Schizophrenien* [*Dementia Precox, or the Group of Schizophrenias*]. Vienna: Franz Deuticke.

Carnap, Rudolf. 1928. *Der logische Aufbau der Welt* [*The Logical Construction of the World*]. Berlin: Weltkreis-Verlag.

Feuerbach, Ludwig. 1948. *Das Wesen des Christentums* [*The Essence of Christianity*]. Leipzig: Otto Wigand.

Freud, Sigmund. 1927. *Die Zukunft einer Illusion* [*The Future of an Illusion*]. London: Hogarth Press.

Galimberti, Umberto. 1999. *Psiche E techne: L'uomo nell'età Della tecnica* [*Psyche and Techne: Man in the Age of Technique*]. Feltrinelli Editore.

Habermas, Jürgen. 1981. *Theorie des kommunikativen Handelns*. Vol. 1: *Handlungsrationalität und gesellschaftliche Rationalisierung*. Vol. 2: *Zur Kritik der funktionalistischen Vernunft* [*The Theory of Communicative Action*. Vol. 1: *Reason and the Rationalization of Society*. Vol. 2: *Lifeworld and System*]. Frankfurt am Main: Suhrkamp.

Hegel, Georg Wilhelm Friedrich. 2018 [1807]. *Phänomenologie des Geistes* [*The Phenomenology of Spirit*]. Oxford: Oxford University Press.

Jaeggi, Rahel. 2014. *Alienation*. New York: Columbia University Press.

Laing, Ronald David. 1959. *The Divided Self: An Existential Study in Sanity and Madness*. Harmondsworth: Penguin.

Laing, Ronald David. 1967. *The Politics of Experience and the Bird of Paradise*. Harmondsworth: Penguin.

Marcuse, Herbert. 1964. *One-Dimensional Man*. New York: Beacon.

Marx, Karl. 1975 [1843]. "Contribution of Hegel's Critique of Right: Introduction". In *Collected Works*, Vol. 3. New York/London: International Publishers.

Marx, Karl. 1975 [1844]. "Manuscripts". In *Collected Works*, Vol. 3. New York/London: International Publishers.

Popper, Karl. 1959. *The Logic of Scientific Discovery*. Abingdon-on-Thames: Routledge.

Prigogine, Ilya. 1978. *From Being to Becoming. Time and Complexity in the Physical Sciences*. San Francisco: Freeman.

Severino, Emanuele. 1979a. *Legge e caso* [*Law and chance*]. Milan: Adelphi.

Severino, Emanuele. 1979b. *Techne: Le radici della violenza* [*Techne: The Roots of Biolence*]. Milan: Rusconi.

Severino, Emanuele. 1981. *La struttura originaria* [*The Originary Structure*]. Milan: Adelphi (1st edition 1958).

Severino, Emanuele. 1982. *Essenza del nichilismo* [*The Essence of Nihilism*]. Milan: Adelphi (1st edition 1972).

Severino, Emanuele. 1990. *Il nulla e la poesia. Alla fine dell'età della tecnica: Leopardi* [*Nothingness and the Poetry. At the End of the Age of Technique: Leopardi*]. Milan: Rizzoli.

Severino, Emanuele. 1996. *La filosofia dai Greci al nostro tempo: La filosofia contemporanea* [*Philosophy from the Greeks to Our Time: Contemporary Philosophy*]. Milan: Rizzoli.

Severino, Emanuele. 1999. *L'anello del ritorno* [*The Ring of the Return*]. Milan: Adelphi.

Severino, Emanuele. 2015. *The Essence of Nihilism*. Edited by Alessandro Carrera and Ines Testoni. London/New York: Verso.

Severino, Emanuele. 2023. *Law and Chance*. Edited by Giulio Goggi, Damiano Sacco, and Ines Testoni. London/New York: Bloomsbury.

Watzlawick, Paul, Janet Beavin Bavelas, and Donald deAvila Jackson. 2011. *Pragmatics of Human Communication: A Study of Interactional Patterns, Pathologies and Paradoxes*. New York: Norton.

Testoni, Ines. 2019. “Eternity between a Novel Theology and a New Science from Giordano Bruno to Emanuele Severino”. *European Journal of Science and Theology* 15(5): 117–132.
Testoni, Ines. 2021. *Il grande libro della morte* [*The Great Book of Death*]. Milan: Il saggiatore.
Testoni, Ines, Dorella Ancona, and Lucia Ronconi. 2015. “The Ontological Representation of Death”. *OMEGA-Journal of Death and Dying* 71(1): 60–81. https://doi.org/10.1177/0030222814568289.
Testoni, Ines, Enrico Facco, and Federico Perelda. 2017. “Toward a New Eternalist Paradigm for Afterlife Studies: The Case of the Near-Death Experiences Argument”. *World Futures* 73(7): 442–456. http://dx.doi.org/10.1080/02604027.2017.1357935.

Names

https://doi.org/10.1515/9783111313610-025

Concepts

https://doi.org/10.1515/9783111313610-026